The Creation of Markets for Ecosystem Services in the United States

Anthem Environment and Sustainability Initiative (AESI)

The Anthem Environment and Sustainability Initiative (AESI) seeks to push the frontiers of scholarship while simultaneously offering prescriptive and programmatic advice to policymakers and practitioners around the world. The programme publishes research monographs, professional and major reference works, upper-level textbooks and general interest titles. Professor Lawrence Susskind, as General Editor of AESI, oversees the below book series, each with its own series editor and an editorial board featuring scholars, practitioners and business experts keen to link theory and practice.

Anthem Strategies for Sustainable Development Series
Series Editor: Professor Lawrence Susskind (MIT)

Anthem Climate Change and Policy Series
Series Editor: Dr. Brooke Hemming (US EPA)

Anthem Diplomacy at the Food-Water-Energy Nexus Series
Series Editor: Professor Shafi qul Islam (Tufts University)

Anthem International Environmental Policy Series
Series Editor: Professor Saleem Ali (University of Delaware)

Anthem Big Data and Sustainable Cities Series
Series Editor: Sarah Williams (MIT)

Included within the AESI is the Anthem EnviroExperts Review. Through this online micro-review site, Anthem Press seeks to build a community of practice involving scientists, policy analysts and activists committed to creating a clearer and deeper understanding of how ecological systems – at every level – operate, and how they have been damaged by unsustainable development. This site publishes short reviews of important books or reports in the environmental field, broadly defined. Visit the site: www.anthemenviroexperts.com.

The Creation of Markets for Ecosystem Services in the United States

The Challenge of Trading Places

Mattijs van Maasakkers

ANTHEM PRESS

Anthem Press
An imprint of Wimbledon Publishing Company
www.anthempress.com

This edition first published in UK and USA 2019
by ANTHEM PRESS
75–76 Blackfriars Road, London SE1 8HA, UK
or PO Box 9779, London SW19 7ZG, UK
and
244 Madison Ave #116, New York, NY 10016, USA

First published in the UK and USA by Anthem Press 2016

British Library Cataloguing-in-Publication Data
A catalogue record for this book is available from the British Library.

ISBN-13: 978-1-78527-123-6 (Pbk)
ISBN-10: 1-78527-123-7 (Pbk)

This title is also available as an e-book.

In memory of Lukas van Maasakkers

CONTENTS

ILLUSTRATIONS

Figures

Tables

ACKNOWLEDGMENTS

This book is the outcome of multiple displacements, professionally and personally. It started as a dissertation at MIT's Department of Urban Studies and Planning (DUSP), expanded during a postdoctoral fellowship at Harvard's Kennedy School of Government and ultimately finished at Ohio State University's City and Regional Planning section. Along the way, a significant number of people contributed to the manuscript through feedback, conversations and brainstorms. None of them contributed more significantly than my dissertation committee members, Larry Susskind, Sheila Jasanoff and the late Judy Layzer. It is hard to imagine a more supportive, engaged and stimulating set of professors to shepherd one through the dissertation process and beyond. Larry and Sheila continue to be wonderful mentors, for which I am deeply grateful. Judy's untimely passing is a deeply felt loss to the community of scholars on environmental policy and planning, and I personally miss her good humor, relentless focus on writing style and quality and incisive feedback.

A project like this relies on the willingness of busy practitioners to explain and discuss their work with an academic outsider like me. For their generosity and patience I would like to sincerely thank all the people who volunteered their time to answer my questions, in person, over the phone or in writing. Their insights and passion have made my exploration of markets for ecosystem services truly enjoyable, engaging and frequently surprising.

Many friends, colleagues and (former) students have contributed to this book in a myriad of ways. I would like to thank all of them for listening patiently while I was going on and on about the latest version of the Shade-a-Lator or some other element of the project that they probably didn't really care to hear about (again). At MIT, I benefited tremendously from my conversations with fellow DUSP PhD students in my cohort as well as Todd Schenk, Atul Pokharel and Nick Marantz, Danya Rumore, Kelly Heber and Bruno Verdini. Throughout my time as a graduate student I enjoyed support from TNO, the Netherlands Organization for Applied Scientific Research. Adriaan Slob, Mike Duijn and Jos Brils (now with Deltares) in particular were always willing to listen to my halting explanations of profoundly foreign policies and practices. At Harvard, I enjoyed numerous wonderful conversations about this project with the other fellows in the Science, Technology and Society program at the Kennedy School and during my time in the Belfer Center. At Ohio State, I have found a wonderful community of scholars in and around the Knowlton School. Moving to the Midwest made it especially rewarding to study the efforts in that region, and I would like to thank the students who supported me during that phase of the project: Alex Wesaw, Emily Knox, Catherine Brokenshire and David Kilroy.

Finally, I would like to thank my family in the Netherlands and the United States for their support and patience throughout graduate school and now beyond. My wife, Sarah, introduced me to planning and has continued to teach me about this wonderful field. She makes me want to be a better person every day, for which I am deeply grateful. My sons, Frits and Menno, have enlightened and enlivened every day of my life since their births.

These people have helped in the creation of this book tremendously, but all the mistakes remaining, if any, are mine.

Chapter 1

INTRODUCING ECOSYSTEMS
TO THE MARKETPLACE

On Sunday, September 7, 2008, the federal government took control of Fannie Mae and Freddie Mac, which was a dramatic moment in the financial crisis that originated in the housing market.[1] Estimates of the cost to the government of this extraordinary action were around $25 billion.[2] This event, and the broader financial crisis it was a part of, made longtime proponents of the markets that had just collapsed question the most basic assumptions underlying those institutions.[3]

Three days after this takeover, a group of people gathered in Ellicott City, Maryland, a short drive from Washington, DC, to discuss the creation and expansion of a very different kind of market. The official title of the event was the "Market-Based Conservation Incentives Workshop," but in the following years, this now annual gathering was renamed the "Ecosystem Markets Conference." This first meeting brought together more than 100 participants from diverse groups, including environmental advocates from the Environmental Defense Fund and Defenders of Wildlife, government officials from the U.S. Department of Agriculture and the Fish and Wildlife Service, as well as private forestry managers and landowners. The description of the first session outlines the basic premise of the workshop: "Ecosystem services

---

1 The Federal National Mortgage Association and the Federal Home Loan Mortgage Corporation, as Fannie Mae and Freddie Mac are officially known, were important in the expansion of the market in mortgage-backed securities. For an accessible analysis of the role of these institutions in the financial crisis, see: http://hereandnow.wbur.org/2012/01/12/fannie-freddie-financial. Last accessed on 07/30/2013. Ecosystem services credits share only the broadest elements with mortgage-backed securities, but the takeover of these two financial institutions was at the heart of a profound discussion about basic tenets of economic thinking like rational behavior and market actors' abilities to protect their interests.

2 This estimate was published by the Congressional Budget Office, and considered low by some. See: Labaton, S. and Andrews, E. "In Rescue to Stabilize Lending, U.S. Takes Over Mortgage Finance Titans," in the *New York Times* on September 7, 2008.

3 Probably the most widely known example of this kind of questioning is Alan Greenspan's testimony to the U.S. House of Representatives on October 23, 2008, but he was certainly not the only one. The full text is available at: https://house.resource.org/110/org.c-span.281958-1.pdf. Last accessed on 07/30/2013.

provided by working forests—such as clean water, species habitat, and carbon sequestration—have always been undervalued. Ecosystem markets are increasingly recognized as a way to capture the value of these benefits and address natural resource issues."[4] It is clear from the program and the informal reports of some of the participants that creating markets for ecosystem services was a broadly shared ambition. The goal of the workshop was neatly summarized by the title of the second day's sessions: "Moving Ecosystem Markets Forward." Participants discussed successful examples of ecosystem markets, but mainly focused on efforts to expand them and create new markets.

This book is about efforts to develop ecosystem service markets (ESMs) in the United States, and more specifically about why it has proven very difficult to do so successfully. But before trying to understand the challenges associated with creating these markets, we have to address the basic question of why someone would even try. The enthusiasm regarding markets for ecosystem services reflected in this workshop is striking given the broader context in which the meeting was taking place. So why did the participants of the ecosystem market workshop come together to discuss the creation of a whole new category of market institutions at the very moment when conventional markets were undergoing such a dramatic failure?

1.a The Promise(s) of Markets for Ecosystem Services

To understand the appeal of ESMs it is helpful to begin with a deeper look at both practical and theoretical considerations involved in creating such markets. Markets for ecosystem services are considered promising for three major reasons. The first is that for many organizations and individuals involved in environmental issues, including those represented at this workshop, ESMs have the potential to solve very important practical problems in environmental management. The second reason is that the theoretical underpinnings of the calls for this new kind of institution are both deeply entrenched and increasingly prominent in economic scholarship. The third reason is that at the time of this workshop, some nascent attempts to create markets for ecosystem services showed promise in meeting the practical and theoretical expectations described later in this volume.

The basic elements

The diversity of definitions of what a market for ecosystem services is or should be means that the participants in this workshop had varied ideas about

---

4 The title of this session was "The State and Trends of Ecosystem Markets," and it was hosted by Katherine Hamilton, the director of Ecosystem Marketplace, a project dedicated to providing information on markets for ecosystem services.

precisely what they were trying to "move forward." Ecosystem services can be defined as "the conditions and processes through which ecosystems, and the species that make them up, sustain and fulfill human life" (Daily 1997, 3). Following that definition, creating *markets* for ecosystem services provides a new way in which environmental protection and restoration can be organized, measured and incentivized. Different approaches to buying, selling and measuring ecosystem services have been developed, but each one of these activities provides some insight into the popularity of the idea of creating markets for ecosystem services for particular groups.

In most versions of the market idea, the sellers of ecosystem services are landowners. So for a tree farmer attending the ecosystem markets workshop, the creation of new markets represents the potential for a new stream of revenue. Traditionally, forestry management has focused on producing revenue through the sale of timber. However, while those trees are growing they can also provide a lot of additional "services," like capturing carbon dioxide and reducing erosion by fixing soils. If certain individuals or organizations are willing or required to invest in these services, that tree farmer might be able to gain additional income. This might seem like a far-fetched idea, but there is an oft-repeated example, which was presented at the 2008 ecosystem markets workshop in Maryland, of this type of payment.[5] This is the so-called Catskills story. While the facts underlying this case are somewhat disputed (Glenna 2010; Sagoff 2002), it is typically recounted as a successful effort by the City of New York to fund land conservation and restoration around the sources of its water supply, far outside the city limits in the Catskills Mountains, to protect the quality of its drinking water. It is easy to see why this story would be appealing to landowners, since the quality of urban water supply is a widespread concern, and the changes required in land management practices to receive those payments are possibly quite minimal.

The idea of a big city paying for land conservation and restoration far outside of its own jurisdiction to avoid the costs of additional mechanical filtration explains some of the appeal of ESMs for environmental advocates. Many environmental organizations buy land to protect or restore landscapes, and the potential for local governments and other organizations to contribute to environmental conservation and restoration is highly attractive to those entities. The prospect of cities protecting their interests by investing in the environment makes it easy to imagine many more restoration and conservation projects being implemented. Adding urban infrastructure budgets as possible sources of funding for ecological restoration would result in much more

---

5 The executive director of the Watershed Agricultural Council, Tom O'Brien, gave a presentation titled: "A Watershed Model for Water Quality Services and Rural Economies in New York."

available funding than exclusively relying on philanthropy and federal grants. Cities or private companies trying to protect the quality of their water supplies or the level of flood protection they enjoy are additional hoped-for buyers of credits for forests or wetlands in the view of some environmental groups.

The second reason ESMs appear particularly promising for environmental advocates, besides the hope of making more money available for environmental protection, is that a functioning market can provide a powerful brake on the development of rural lands. Many private rural landowners, like farmers and foresters, are facing a constant choice between continuing their work and selling their land to real estate developers who will turn farms and forests into shopping malls or suburbs. An additional steady stream of revenue, separate from commodities markets for grains or timber, might convince some of those landowners that farming continues to be preferable over selling their land to developers.[6] A closely related argument for environmentalists is that the existence of this type of market mechanism will help advocacy organizations reach out to private landowners to convince them that environmental stewardship is a worthwhile endeavor (Bean et al. 2003).

The government representatives at the ESM workshop in Maryland came primarily from the Department of Agriculture, but also from the Environmental Protection Agency and the Department of the Interior. Given the different roles and tasks of these organizations, their official views on ESMs vary.[7] The Department of Agriculture was directly instructed by Congress to work on the development of markets for ecosystem services. The Food, Conservation and Energy Act of 2008, generally referred to as the Farm Bill, included section 1245, titled "Ecosystem Services Markets." This section requires the secretary of agriculture to promote and develop a range of ESMs. In response, Agriculture created the Office of Ecosystem Services Markets. Now known as the Office of Environmental Markets (OEM), this unit focuses on and promotes the precise measurement of ecosystem services production. The potential for

6 These broad reasons for the enthusiasm about markets for ecosystem services among environmental advocates can be found in some form in a variety of books and reports (Majanen, Friedman and Milder 2011; Scarlett and Boyd 2011). At the ecosystem markets workshop in 2008, Robert Bonnie of the Environmental Defense Fund gave the opening address, and his expression of these arguments can be found as early as 2003, when he coauthored a report titled: "The Private Lands Opportunity: The Case for Conservation Incentives (Environmental Defense)."

7 To get a detailed sense of some of the differences in emphasis, it is interesting to compare the three factsheets these departments produced on a single, relatively well-known ecosystem services project, the Clean Water Services Tualatin shade-trading program: The Department of the Interior's Fish and Wildlife Service factsheet can be found at: http://www.fws.gov/oregonfwo/LandAndWater/Documents/ESMflyer_Dec2011.pdf, and the Department of Agriculture's is located online at: http://www.fs.fed.us/ecosystemservices/pdf/Watershed_Services.pdf. An EPA factsheet on the Tualatin program can be found at: http://www.epa.gov/npdes/pubs/wq_casestudy_factsht4.pdf. All last accessed on 05/19/2015.

precise and specific measurement of the environmental performance of land conservation and restoration is regularly invoked as a benefit of market creation, and the Office for Environmental Markets is actively involved in the development of several of these measurement systems. Generating and selling credits based on environmental restoration involves the quantification of the specific volume of a service that has been produced. For example, rather than simply saying that one tree equals a carbon credit, markets for carbon dioxide rely on detailed measurement systems that convert variables like the width and height of planted trees to a volume of carbon dioxide "captured" as a result of their growth (García-Oliva and Masera 2004).

The thought of expanding this sort of detailed accounting to other types of ecosystems and the movement to include services beyond capturing greenhouse gasses have gained popularity, and not just inside of the Department of Agriculture (Stephenson and Bosch 2003). The Environmental Protection Agency convened a series of meetings on ecosystem valuation and its role in decision making as early as 1991. These meetings were organized "for the purpose of advancing the state of the art of ecosystem valuation methods" (Bingham et al. 1995) and included prominent researchers in this field, like Robert Costanza and Richard Norgaard. Scholars and government officials consider such measurement tools a way to include a more complete accounting of the value of the environment in natural resource damage assessments, environmental impact assessments and cost-benefit analyses (Salzman 1997; National Research Council (U.S.) 2005).

The most basic elements of markets for ecosystem services, namely measuring, buying and selling credits, hold different kinds of promises for specific groups and organizations. This does not mean that the ESM workshop was simply an exercise in wishful thinking. The theoretical foundations that form the basis for these hopes and expectations are well established, yet also not completely uniform.

Environmental markets and payments for ecosystem services

ESMs did not suddenly appear as a brand new idea in the early 2000s. Despite their popularity, and some agreement about their most basic elements, there is no clear consensus about the exact definition of what a market for ecosystem services is or should be. The concept has roots in at least two fields of scholarship, namely, environmental economics and ecological economics. The similarities and differences between these two fields is itself a topic of academic study (Munda 1997), but together they provide much of the theoretical foundations for the tools, approaches and expectations related to the creation of markets for ecosystem services. The foundational work in environmental economics in the 1960s and 1970s focused on the creation of markets and property rights regimes to deal with issues of pollution, typically from readily identifiable sources like factories and water treatment plants. Ecological economics, which rose to prominence decades later in the 1990s, has taken the production of benefits through ecosystem functions and processes as its starting point. These two ideas are not mutually exclusive, but have resulted in a situation where the precise theoretical underpinnings of the expectations about markets for ecosystem services can be hard to pinpoint.

The market is a foundational concept in modern economics, and its basic definition and underlying assumptions have been debated since Adam Smith's *Wealth of Nations* (1999 [1776]). The use of markets is often associated with two important goals in neoclassical environmental economics: minimizing the overall costs of reducing environmental pollution and encouraging technological innovation. The idea of achieving environmental protection by deliberately creating markets dates back to at least 1968, when economist John Dales published a brief text titled *Pollution, Property and Prices*. This now classic essay laid out the idea of creating a market in "pollution rights" as an efficient way of reducing pollution without the need for "heavy-handed" government intervention. In the decades following Dales' publication, his idea gathered academic (Ackerman and Stewart 1987) and regulatory support (Freeman and Kolstad 2007). A number of markets in "pollution rights" were developed, the earliest focusing on air pollution (Ellerman 2000).

Starting in the early 1990s, multiple "pollution rights" markets and similar arrangements have been established to achieve environmental goals. The popularity of these market-based instruments provides a strong basis from which to argue that neoclassical economic theory can be harnessed in concrete ways to achieve environmental goals. The sulfur dioxide emissions trading scheme under the Clean Air Act probably remains the prototypical environmental market in the United States following the principles Dales had laid out (Stavins 1998; Tietenberg and Lewis 2012). Reduced to its most basic components, this market relies on the government first establishing an emissions standard for a pollutant, in this case sulfur dioxide, which causes acid rain. Factories discharging that pollutant are then made aware of the standard. When a factory reduces its emission of the pollutant, for example by installing a filter, it can receive *emission reduction credits* for that action. If a different factory exceeds the standard, because of an increase in its emissions resulting from an expansion or when the legal emissions standard is made more stringent, it can purchase the credits from the factory that reduced its emissions to compensate for going over the standard. Therefore, reducing pollution below a facility's allocated amount produces the equivalent of money in the bank, with credits available for sale when the market produces opportunities.

The key benefit of this mechanism is a clear incentive to reduce pollution, which in turn is likely to promote new and potentially cheaper emissions reduction technologies (Tietenberg 1990). However, this kind of environmental market requires an enforceable standard set by a government agency, based on regulation, and is largely agnostic when it comes to the type of technology used to secure the emissions reduction, since innovation in technology is expected.[8] Accordingly, much of the enthusiasm and many of the goals for markets for ecosystem services, for example to encourage the protection of existing ecosystems, do not immediately emerge from this type of institution.

The second set of theories and expectations about markets for ecosystem services derives from the popularization of the ecosystem services concept in the late 1990s (Costanza et al. 1997; Daily 1997), although its origins can be traced to earlier academic work (Westman 1977). Ecological economics generally emphasizes the value of existing ecosystems to human survival and presumes that it is possible to measure the production of specific services by (parts of) ecosystems and landscapes. An important publication in that context is *Nature's Services: Societal Dependence on Natural Ecosystems*. Gretchen Daily, the book's editor, describes the origins of these ideas in a meeting of a group of Pew Fellows in Conservation and the Environment around 1993. She writes in the preface: "A small group gathered informally to lament the near total lack of public appreciation of societal dependence upon natural ecosystems" (1997, xv). One of the members of that group, Robert Costanza, was also the lead author of another 1997 publication, an article titled "The Value of the World's Ecosystem Services and Natural Capital," published in the journal *Nature*. Costanza and his coauthors estimate the total average value of global ecosystems services at approximately $33 trillion annually, providing a compelling economic argument for the protection of those ecosystems (1997, 259). This article has been cited a whopping 13,458 times, attesting to its huge influence.[9] The central insight of both of these popular works is the notion that existing ecosystems provide specific benefits to human beings and societies.

It is hard to overstate how popular and widespread the concept of ecosystem services has become. The ecosystem services concept, and the subdiscipline dedicated to studying it, namely *ecological economics*, rapidly gained

---

8 There is environmental economic scholarship on environmental markets in the absence of regulatory limits as well, sometimes referred to as *Free Market Environmentalism*, which is also the title of a seminal text in this field (Anderson and Leal 1991). Much of this work is associated with the Property and Environment Research Center (PERC). See: http://perc.org. Last accessed on 05/11/2015.

9 The number of citations is clearly a crude indicator of an article's influence or importance, but this very large number of citations, also in comparison to almost any other publication in *Nature*, does give a sense of the popularity of the ecosystem services concept. This specific number was found using scholar.google.com on 05/11/2015.

acceptance in academic circles, evidenced by the creation of professorships,[10] journals,[11] and departments and programs at universities.[12] The number of academic publications mentioning this concept increased exponentially over the next 15 years (Fisher 2009). Outside of academia, a number of large international research projects that explicitly focus on ecosystem services have been created, like the UN-initiated Millennium Ecosystem Assessment (or MA, www.maweb.org) and The Economics of Ecosystems and Biodiversity (or TEEB, www.teebweb.org). These projects involve thousands of academics and other researchers and have resulted in a vast number of publications on the state of ecosystems and the goods and services they provide around the world. The wide circulation of the idea is likely to have contributed in large measure to the enthusiasm and optimism among the people at the Maryland ecosystem markets workshop.

The popularity of the concept shifted attention from *whether* markets should be created to *how* they should be implemented: "Around the world, leaders are increasingly recognizing ecosystems as natural capital assets that supply life-support services of tremendous value. The challenge is to turn this recognition into incentives and institutions that will guide wise investments in natural capital, on a large scale" (Daily and Matson 2008, 9455). The literature on ecosystem services began to focus on implementation through the creation of payments and markets for ecosystem services, rather than more abstract valuation studies (Gómez-Baggethun et al. 2010). A widely used definition of the payments for ecosystem services (PES) approach is: "A voluntary transaction in which a well-defined environmental service (or a land use likely to secure that service) is bought by (a minimum of) one buyer from (a minimum of) one provider if and only if the provider continuously secures the provision of the service" (Wunder 2005, 3).

Some important scholars within the field of ecological economics (cf. Mayrand and Paquin 2004; Sommerville, Jones and Milner-Gulland 2009) consider this vision of PES too closely connected to neoclassical environmental

---

10 For example at the University of Bayreuth in Germany, see: http://www.pes.uni-bayreuth.de/en/publications/topics/index.html. Last accessed on 05/11/2015.

11 Besides *Ecological Economics*, the flagship journal of the International Society for Ecological Economists since its creation in 1989, the *International Journal of Biodiversity Science and Management* was renamed the *International Journal of Biodiversity Science, Ecosystem Services and Management*, and Elsevier has published a new journal titled *Ecosystem Services* since 2012.

12 Prominent examples are the Gund Institute for Ecological Economics at the University of Vermont, the Ecosystem Services for Urbanizing Regions program at Portland State University and the Natural Capital Project at Stanford University. Many other programs are less prominent, but often explicitly connected to ecosystem services. These can be found at: http://www.ecoeco.org/content/education/graduate-opportunities/ and http://www.aashe.org/resources/academic-programs/discipline/economics both last accessed on 05/14/2015.

economics, as opposed to ecological economics (Farley and Costanza 2010; Muradian et al. 2010). Despite these scholarly disputes, this definition highlights the need to define a specific environmental service or land use, which is one element that sets PES apart from the traditional environmental market. The focus is on the ability of an ecosystem to produce outcomes desirable for human beings, as opposed to humans' ability to reduce their adverse impacts on the environment (e.g., pollution). Precise definitions of markets for ecosystem services are even more difficult to find, and several well-known scholars simply use "markets for ecosystem services" as a catchall phrase that covers a variety of schemes including PES, pollution credit trading and sometimes even agricultural subsidies (Achterman and Mauger 2010; Salzman 2005; Womble and Doyle 2012).

The fields of environmental and ecological economics provide a diverse set of theoretical foundations for the impetus to create ESMs. Salient, and some would say fundamental (Munda 1997) differences exist between these two academic disciplines. For example, the foregoing definition of payments for ecosystem services emphasizes the voluntary nature of the transaction, whereas much of the work on pollution trading in environmental economics assumes a regulatory standard to initiate trading. These kinds of theoretical arguments and disagreements do not automatically diminish the appeal of the idea of markets for ecosystem services. It's perhaps more useful to think of *markets for ecosystem services* as a hybrid concept that combines environmental and ecological economics, and the central preoccupations within those fields, in new, sometimes uncomfortable ways. In practice, descriptions and definitions of markets for ecosystem services vary, but some of the key preoccupations described here, such as achieving efficiency through credit trading and the precise measurement of environmental attributes, typically return again and again.

Early success and emerging challenges

The third major reason for confidence in ESMs in workshops and meetings like the Ecosystem Markets Conference is that tangible examples display some of the promises of markets for ecosystem services being fulfilled. One prominent example is in the Tualatin River watershed in Oregon where, as is the case in much of the Pacific Northwest, multiple species of salmon are struggling for survival. This is due at least in part to significant increases in water temperatures over the past decades. Warm water adversely affects survival rates of spawning and smelting salmon (Richter and Kolmes 2005). The conventional method to cool wastewater after it has been treated is to construct mechanical "chillers," which are expensive both to build and to maintain. The water resources utility in the Tualatin River watershed, named Clean Water

Services (CWS), instead developed a water cooling program that includes payments to landowners who plant riparian trees. The ecosystem service these trees provide is shade, or more specifically temperature reduction in the water over which the trees cast shade, which in turn is expected to increase salmon survival rates.

Shade trading, as it is sometimes called, combines several elements found in the literatures on markets and ecosystem services. This specific program has been described extensively in academic journals (Achterman and Mauger 2010; Cochran and Logue 2011). Shade trading has also been approvingly mentioned in policy reports about ecosystem services from national environmental organizations like Resources for the Future (Scarlett and Boyd 2011, 7) and Forest Trends (Stanton et al. 2010, 69). The regulatory impetus for the shade credit program means it does not fall under the definition of a PES scheme as advanced by some scholars,[13] but this has not prevented others from arguing, "The CWS permit epitomizes the promise of ecosystem service markets" (Achterman and Mauger 2010, 304).

This program is the basis for much enthusiasm about creating markets for ecosystem services, but it also reveals some of the challenges of making markets work in practice. These challenges are the core subject of this book. Those that the Clean Water Services overcame in the development of this program can be categorized under three broad rubrics, which help structure the arguments advanced in the remaining chapters. The first was that the logic of markets is at odds with the central role of place in the way many people experience their environment. A tree that is planted next to a stream in a city park might be inconsequential for its effect on water temperature, whereas one planted by a stream running through farmland may produce economically beneficial cooling. But choosing to fund planting one over the other has a range of noneconomic effects as well. It is difficult yet important to incorporate those considerations into the construction of ESMs. The second challenge that was overcome in the Tualatin case but is more problematic in general is developing a measurement system that performs the calculation of the credits, a complicated and often controversial task. The third and final challenge was that a lot of people were involved in or affected by the creation of this market in different ways. Reaching a basic level of agreement, or at least avoiding explicit disapproval expressed through litigation, proved difficult.

Clean Water Services overcame the first challenge with the decision to target its tree-planting efforts in a very deliberate fashion, taking into account a range of considerations, not purely market-based ones. CWS developed a detailed temperature management plan consisting of several elements that

---

13 The most obvious and influential example is the definition cited earlier (Wunder 2005).

mix voluntary and regulatory approaches, as well as infrastructural and ecosystem-based water temperature reduction strategies. The element that is most commonly highlighted is the tree-planting program offered to landowners in the watershed. A second feature of the CWS approach is a tree-planting program in urban areas, the so-called Community Tree Planting Challenge. Outlined in the Clean Water Services' *Healthy Streams Initiative*,[14] this challenge set a goal of planting 2 million trees over 20 years. This has been executed primarily on public lands with extensive help from volunteers and community groups. Getting volunteers to plant trees in urban parks to satisfy the requirements outlined by a legal permit is a creative solution. It also shows that the realization that ecosystems provide benefits can result in new and productive arrangements and collaborations.

The challenge of developing a measurement system to calculate the number of credits was overcome through the creation of an instrument known as the Shade-a-Lator. This is a spreadsheet-based tool that calculates the shadow produced by a tree-planting project in terms of the reduction in the amount of solar energy entering the water, in kilocalories per day. Information about only a few key variables, such as the width of the stream, its direction in relation to the sun and its depth, need to be entered into the spreadsheet to calculate the number of "credits" produced. The Shade-a-Lator is a fairly simple measurement tool, in the sense that it is easy to gather the necessary data and it only calculates a single value. The most complex element of the Shade-a-Lator is probably the Heat Source model,[15] which simulates the transfer of heat energy to stream temperature. This simulation model was created for a master's thesis at Oregon State University (Boyd 1996) and used by the Oregon Department of Environmental Quality to create the Shade-a-Lator. This combination of existing and new models and calculators is characteristic of many ecosystem service measurement systems, and points to some of the tensions between precision, practicality, innovation and reliability inherent in creating them.

The third and final challenge consists of creating a process to reach agreement about the market's design and functioning with relevant authorities and stakeholders. Engaging and reaching agreement with all relevant stakeholders is incredibly complex and hard to implement successfully. The competing

---

14 The Healthy Streams Plan, which outlines the Community Tree-Planting Initiative, contains a detailed analysis of tree-planting options (p. 30) highlighting opportunities that emphasize public lands and proximity to urban areas. Available at: http://www.cleanwaterservices.org/Content/Documents/Healthy%20Streams%20Plan/Healthy%20Streams%20Plan.pdf. Retrieved on: 05/11/2015.

15 Documentation on the Shade-a-Lator, including comments on the Heat Source model and a detailed description of the development of both, can be found at: http://www.deq.state.or.us/wq/trading/trading.htm. Retrieved on: 05/11/2015.

logics of markets and place, as well as the difficulty of creating measurement systems, make it complicated to figure out who needs to be involved in the new market, when and how. Certain actors, such as permitting entities in a regulatory context, need to agree with the basic design and the specific rules of the market. Others, like landowners, need to actively participate by developing credits on their land to make trading work. Yet another category of actors to keep in mind consists of any group or individual who might oppose the market so vehemently that a legal challenge or other blocking action is likely if they are excluded.

In the Tualatin example, overcoming the threat of opposition required a lengthy negotiation between multiple organizations, persistent intermediary efforts by the relevant state agency and some creativity. There is a regulatory reason why CWS has to reduce the water temperature, stemming from the Clean Water Act. This means that the central features of the shade credit trading scheme are outlined in a 70-page legal document officially called a "National Pollution Discharge Elimination System (NPDES) Watershed-Based Discharge Permit," issued by the Environmental Protection Agency and the Oregon Department of Environmental Quality (ORDEQ).[16] These two agencies therefore needed to provide their explicit approval of the trading program.

The Department of Environmental Quality worked closely with CWS to develop this trading program between 2001 and 2006. The agency created an informal group of stakeholders.[17] At different stages in the development of the trading scheme, Clean Water Services, the Department of Environmental Quality and members of the stakeholder group had deeply conflicting views on the appropriate type and level of involvement of those stakeholders. For example, once the basic requirements for shade-trading programs in the state of Oregon were clear, ORDEQ decided to proceed with Clean Water Services to develop the specific permit for its facilities.

The utility viewed this stage of the process as a bilateral negotiation to be conducted between the regulator and the organization seeking the permit. The members of the stakeholder groups wanted to remain involved and ORDEQ feared resistance from an environmental advocacy organization, the Tualatin Riverkeeper. The agency maintained regular contact with that organization

---

16 For a basic description of the NPDES program, see: http://water.epa.gov/polwaste/npdes/. Last accessed on 05/12/2015. A more extensive description of the history and development of the Clean Water Act and the TMDL program is Houck (2002).

17 The Trading Stakeholder Group as well as the broader process to develop the trading program is described in detail in DEQ's Water Quality Trading Case Study, published in 2007 and available at http://www.deq.state.or.us/wq/trading/docs/wqtradingcasestudy.pdf. Retrieved on: 05/12/2015. Some of the specific challenges associated with the procedural dimension of creating this market are described on pages 8–11.

to keep it on board. The profound disagreements over who should participate at what stage(s) of this negotiation process show that straightforward assumptions about when to involve which stakeholders do not readily translate into the realm of markets for ecosystem services.

In the end, the NPDES permit CWS received in 2005 is widely considered innovative and efficient, I believe rightly so, especially given the potential cost of building mechanical "chillers."[18] However, it does not fulfill all the hopes and expectations for markets for ecosystems services emerging from either practical ideals or economic theory. For one, the final element of the Clean Water Services approach draws on traditional "grey" infrastructure rather than planting trees. When added together, the shade activities have achieved between 10% and 30% of the total temperature reduction stipulated in the NPDES permit for the period 2004–2009.[19] The other 70–90% is achieved by what CWS calls "flow augmentation." That part of the temperature management plan consists of water releases from reservoirs created by dams, to which CWS holds water rights. This more conventional aspect of the Tualatin story is rarely mentioned in the factsheets[20] and publications on markets for ecosystem services (Abdalla 2008).

The Tualatin example, though often relayed in a highly simplified form, was and still is very popular among many of the proponents of markets for ecosystem services. It gave the participants in the ecosystem markets workshop a concrete example in which some of the promises of this new and exciting idea, or complex of ideas, were being achieved. Upon closer examination, however, this story also reveals a set of challenges to the creation of ESMs, and the ways in which Clean Water Services, the Oregon Department of Environmental Quality and the many other participants overcame them.

1.b Why Are Markets for Ecosystem Services so Difficult to Create?

Despite the popularity of the idea of ESMs and the concerted efforts of many to create them across the United States, there are still relatively few active and

---

18 CWS reports annually on the progress of the various elements of its temperature management program via its website: http://www.cleanwaterservices.org/. Last accessed on 05/14/2015.

19 See: CWS Temperature Management Plan Annual Report 2009. Available at: http://www.cleanwaterservices.org/Content/Documents/Projects%20and%20Plans/Temperature%20Management%20Plan%20Annual%20Report%202009.pdf. Last accessed on 05/12/2015.

20 The Department of the Interior's Fish and Wildlife Service factsheet can be found at: http://www.fws.gov/oregonfwo/LandAndWater/Documents/ESMflyer_Dec2011.pdf and the Department of Agriculture's at: http://www.fs.fed.us/ecosystemservices/pdf/Watershed_Services.pdf. An EPA factsheet on the Tualatin program can be found at: http://www.epa.gov/npdes/pubs/wq_casestudy_factsht4.pdf. All accessed on 05/13/2015.

stable markets of this kind. Even fewer of those existing markets meet the goals of their advocates or live up to the theoretical expectations outlined by various economists. Therefore, the basic question that this book seeks to answer is: *Why are markets for ecosystem services so difficult to create?* Of course, making new institutions is always difficult, so before expanding further on this question and my approach to answering it, it is important to understand what is specific to this case by looking in detail at the evidence for the limited expansion of markets for ecosystem services. I will then look at two popular explanations for the limited practical realization of markets for ecosystem services: first, that the economic downturn of 2008 prevented their expansion, and, second, that their limited success is an example of well-known implementation problems that come up in any policy setting.

Failure to launch

In his book *Nature and the Marketplace: Capturing the Value of Ecosystem Services*, Columbia Business School professor Geoffrey Heal asked how much scope there is for "using markets to manage our interactions with the natural environment" (2000, 185). In the years since he posed that question, a group of dedicated people has worked hard to create and expand those markets in the United States. More than a decade later, it is hard to argue we live in "a world in which people and institutions appreciate natural systems as vital assets, recognize the central roles these assets play in supporting human well-being, and routinely incorporate their material and intangible values into decision making" (Daily et al. 2009, 21). Yet the development and expansion of markets for ecosystem services has been slow. Setbacks and failed efforts are quite common, and the most well-known examples remain the Catskills and Tualatin stories.

Relatively few active and stable ESMs exist in the United States. Widely agreed upon counts of how many exist are difficult to come by, in large part because of the lack of a practical definition of that concept. Ecosystem Marketplace, an organization involved in the promotion and expansion of a range of different types of markets for ecosystem services, produces separate analyses for markets[21] primarily related to water (or "payments for watershed services," in their phrase), those focused on species habitat (referred to as "biodiversity markets") and finally on markets related to carbon sequestration.

One prominent type of payments for watershed services is Water Quality Trading, the rubric under which the Tualatin shade credit program falls. Not

---

21 The most recent report on watershed payments is: Bennet, G. and Carroll, N. *Gaining Depth: State of Watershed Payments 2014*, published in 2015 by Forest Trends under its Ecosystem Marketplace project. Similar reports on the state of markets for ecosystem services related to carbon and biodiversity are published by Forest Trends on an irregular basis. See: http://www.forest-trends.org/publications.php. Last accessed on 05/13/2015.

all of the Water Quality Trading programs in the United States allow for the creation of credits through environmental restoration like tree planting. Many of these trading programs primarily consist of trades between water treatment facilities or other so-called point sources as a result of technological upgrades in filtration capacity, following the logic of pollution credit trading. Despite the enthusiasm about the Tualatin shade credit program, Water Quality Trading itself was far from new in 2005. Early examples, both between existing water dischargers and including non-point sources like farmers, can be traced back to the early 1980s (McGinnis 2001). The EPA started supporting the creation of trading programs to achieve water quality goals as early as 1996, when the agency issued its Draft Framework for Watershed-Based Trading,[22] which was followed by its Final Water Quality Trading Policy[23] in 2003. However, the development in the overall number of Water Quality Trading programs of this type over time reveals a broad trend related to markets for ecosystem services, namely that many efforts to create them are initiated, few succeed, and even fewer remain active for an extended period of time.

In 2008, Ecosystem Marketplace counted 66 Water Quality Trading programs, or efforts to create such a program, in the United States. Forty-five of those programs identified had either never been active (31) or no information could be found about them (14). Of the remaining 27, only 12 had seen actual credit trades occur in 2008, meaning that the other 15 had either collapsed permanently or simply did not have any activity that year (Stanton et al. 2010, 56). This analysis also included a look at the development in the number of Water Quality Trading programs over time, and found that the number of programs that had seen trading activity before 2000 was 13, and that between 2000 and 2008 the average number of active programs was between 11 and 12. In 2013, this analysis was updated using slightly different methods, and 23 active Water Quality Trading programs were counted, almost doubling the number found in 2008 (Bennett, Carroll and Hamilton 2013, 58). Upon closer inspection, that same report shows that only 15 programs actually registered trading activity in 2011, the most recent year for which information was available. In 2013, a separate analysis counted a total of 35 Water Quality Trading schemes in the United States that were connected to regulatory requirements (NPDES permits), and found detailed information about 18 of those (EPRI 2013). While credit trading was allowed under all of these permits, only 11 of the programs actually applied credits to meeting the permit requirements.

---

22 Available for download here: http://epa.gov/owow/watershed/framework.pdf. Retrieved on: 05/13/2015.

23 This policy can be found in its entirety at: http://water.epa.gov/type/watersheds/trading/final-policy2003.cfm. Last accessed on 05/13/2015.

Another way to compare the overall scale of Water Quality Trading is to look at the transaction volume in terms of the actual money spent, and in 2008 the total dollar value of credits transacted in North America was $10.7 million. In 2011 the value of credits transacted was $7.7 million, showing that the total value of financial activity in water quality markets decreased between those years (Bennett, Carroll and Hamilton 2013, 59). A third analysis, published in 2015, shows that the trading volume in 2013 had reached $11.1 million and indicated that the longest-running markets are decreasing in volume, meaning fewer trades are taking place while credit prices are stable or only slightly increasing (Bennett and Carroll 2014, 83–84).

These numbers are influenced by overall economic activity, and perhaps even the achievement of some of the efficiency goals associated with market creation, assuming individual credit prices were dropping in some markets. But more broadly, they reflect the simple conclusion that markets for ecosystem services, while remaining popular in theory, are not expanding or growing significantly. A separate, broader review of the implementation of ecosystem services programs in the United States and Canada concluded: "Many of the PES schemes are still in the 'proof-of-concept' stage, with initial or annual payments from the government or other parties needed to instigate changes in behavior and land management practices" (Molnar and Kubiszewski 2012, 53).

A final example of the limited expansion of markets for ecosystem services is the shade credit trading program in the Tualatin River watershed itself. So far, only one additional city in Oregon has received its own NPDES permit that includes shade trading.[24] A few other cities continue to consider the shade credit approach, but in 2014 the Oregon Department of Environmental Quality announced it was writing new rules for Water Quality Trading in the state. This announcement, made after severe criticism of Oregon's trading policy became public, has effectively halted the issuance of new permits that allow credit trading in Oregon. A decade after Clean Water Services received its permit, this approach has not expanded widely despite consistent and well-funded efforts to promote it. All of this makes clear that the optimism about markets for ecosystem services has not been translated into widespread implementation.

Economics and policy

The realization that markets for ecosystem services have not taken the United States by storm has not escaped the community of people most dedicated to this idea. The Ecosystem Markets Conference merged with a more research-oriented conference on ecosystem services, at least in part as a result

24 See: http://www.deq.state.or.us/wq/trading/trading.htm#Rep. Last accessed on 05/13/2015.

of declining sponsorship. At this newly merged conference, popular sessions included[25] "Taking Ecosystem Service Markets to Scale: What Will It Take?" and "Where Buyers Are Coming to the Market: Moving Beyond Pilot Trades." These sessions reveal that markets apparently have not "gone to scale" (yet), and that many efforts have remained in the pilot project stage, sometimes for years.

In conversations and publications on markets for ecosystem services, two reasons for this stagnation are most commonly given. The first is that the financial crisis and ensuing recession, which coincided with the emergence of ecosystem service markets as a viable approach, drastically reduced the momentum and success of many of the market-making projects. The second is that the implementation of any policy idea is complicated, will take a long time and is pretty likely to fail. These arguments are so widespread that it is worth describing their basic logic in some detail, as I do in the following sections. These accounts describe developments and obstacles that have added to the complexity of making markets for ecosystem services work, but they fail to make sense of the more fundamental challenges embedded in the idea of markets for ecosystem services. In later chapters, I will turn to those more specific challenges, which were recognizable even in the Tualatin success story.

The economic downturn may explain some failures but certainly not all. The basic distinction between voluntary and regulatory markets is important in this regard. In voluntary markets, individuals or organizations decide to invest in ecosystem services because they believe it to be in their interest. It is plausible to argue that the overall decline in investment affected the creation of such markets for ecosystem services. Before the financial crisis, there was a lot of excitement about private investors entering markets for ecosystem services in a significant way. A cover story on payments and markets for ecosystem services in *The Economist* declared, "All these payments and new markets have not gone unnoticed in the City of London, and other financial centres."[26] The most specific example in that article was that "banks such as ABN AMRO plan to start selling new environmental financial products." Also in 2005, Goldman Sachs created its Center for Environmental Markets and declared the organization would "aggressively seek market making and investment opportunities in the environmental markets."[27] After the financial crisis,

---

25 This conference, titled "ACES and Ecosystem Markets 2012," was held in Ft. Lauderdale, Florida, from December 10–14. The conference program and additional information can be found at http://www.conference.ifas.ufl.edu/aces12/index.html. Last accessed on 05/12/2015.

26 See: *The Economist*, "Environmental Economics: Are You Being Served?" 2005. Available online at: http://www.economist.com/node/3886849. Last accessed on 05/12/2015.

27 See: Goldman Sachs's Environmental Policy Framework, available at: http://www.goldmansachs.com/citizenship/environmental-stewardship-and-sustainability/environmental-policy-framework/gs-environmental-policy-framework-pdf.pdf. Retrieved on: 05/12/2015.

this type of talk, let alone action, by major banks entering environmental markets became much more rare. For example, the Dutch government acquired ABN AMRO in 2009, which resulted in a drastic reduction in its activities outside the Netherlands. Goldman Sachs maintains its Environmental Markets Group and the Center for Environmental Markets, but its main commitment to markets has remained to the European Union Emissions Trading Scheme, a regulatory market.

The economic downturn is not as clearly an explanation for stagnation in the creation of markets in which trading is required or enabled by a specific legal standard. For example, in Water Quality Trading, the impetus to develop a market mechanism is regulatory and not directly connected to the overall level of economic activity in a watershed or state. A water utility in Oregon facing a permit renewal that includes a temperature standard might be expected to pursue a trading strategy even more aggressively in economically depressed times, given the cost difference between building a mechanical chiller and purchasing shade credits. The economic downturn might explain low trading volumes in certain existing markets. For example, a water treatment facility might process less water as a result of lower industrial water use in its service area, since production levels in a local factory are down. This could mean that the utility buys fewer credits in a given year, or does not even meet the "cap" and buys none. However, since many of the markets for ecosystem services are driven by existing regulation, the basic establishment of markets is less directly connected to the larger economy. So while the lack of increased trading volumes might be directly related to the recession in the United States, the enduring problems associated with the establishment of new markets for ecosystem services cannot be explained so easily. Moreover, the recession is only relevant to the difficulties of making markets since the beginning of the financial crisis, around 2007. Many of the efforts to create markets, including some analyzed in this book, started years before the downturn and their failure must have other explanations.

The second popular reason for limited success, which has been advanced by prominent legal scholars as well as practitioners, is that the absence of durable implementation is primarily related to the reliance of ESMs on existing laws and policies. These cases fall, in other words, under the well-known heading of implementation failures. This argument is primarily relevant to regulatory markets. In the literature on public policy, the fact that laws and policies are hard to implement is well established (Bardach 1977; Pressman 1984). Of course, the very idea of creating markets as policy tools is in part a reaction to problems of regulatory implementation (Stavins 2003). An example of this kind of reasoning, specific to Water Quality Trading, is the statement that "realizing the potential of market-like approaches will require moving beyond promotional rhetoric and developing workable options for the barriers erected by the interpretation and implementation of regulatory programs called for under the CWA [Clean Water Act]" (Stephenson and Shabman 2011, 16). This line of argument is at odds with the fact that some of the key factors that would lead us to expect successful implementation are present. For one, many important officials have expressed support for this approach,[28] and a group of committed professionals has worked for years to implement it. Significant resources have been leveraged, including millions of dollars in federal grants and subsidies.[29] Second, the EPA has written extensive guidance on specific markets for ecosystem services, like Water Quality Trading, and it is obviously working in some places, like the Tualatin River watershed.[30] Simply pointing to the general implementation problem obscures the more specific, and I will argue fundamental, challenges that emerge when trying to create markets for ecosystem services.

The economic and policy implementation arguments do not adequately explain why motivated and well-funded practitioners have struggled so hard to implement markets, when there is compelling evidence that these markets promise to facilitate cost-effective, environmentally friendly permit compliance. In order to unearth better reasons, I followed a somewhat unconventional research strategy.

1.c How to Study Markets for Ecosystem Services

To explore why it has proven so difficult to create markets for ecosystem services, I chose to look at the practical efforts to create those markets in three regions in the United States: the Chesapeake Bay watershed, the Willamette River basin and the Ohio River basin. The methods I used to analyze markets for ecosystem services, and how I came to select these three locations as the focus for this book, are deeply connected and require some explanation.

The most straightforward reason to focus on practice in a limited number of places is that theoretical expectations for ESMs can be found in many

---

28 An obvious example is the language in the 2008 Farm Bill instructing the USDA to work on the development of these markets. A more recent and perhaps more visible example is a speech by President Obama in 2012, in which he described a shade credit program in Oregon in glowing terms. This speech can be found in its entirety at: http://www.whitehouse.gov/photos-and-video/video/2012/03/02/president-obama-speaks-conference-conservation. Retrieved on: 05/13/2015.

29 Both the EPA and USDA have supported the creation of markets for ecosystem services through grants. The EPA's Targeted Watershed Grant program and especially the USDA's Conservation Innovation Grant program have funded many efforts going back to 2002. In 2012, the USDA funded 12 projects related to Water Quality Trading, at a cost of roughly $7 million. For the EPA program, see: http://water.epa.gov/grants_funding/twg/initiative_index.cfm and the USDA: http://www.nrcs.usda.gov/wps/portal/nrcs/main/national/programs/financial/cig/. Last accessed on 05/12/2015.

30 See: http://water.epa.gov/type/watersheds/trading.cfm. Last accessed on 05/12/2015.

bodies of thought and are not uniform. Making a market for ecosystem services is not simply an economic activity for which the main challenges can be explained by theories about externalities; nor is it purely an ecological activity where the most relevant insights can be drawn from theories about appropriate habitats for an endangered species or other such phenomena. Rather than starting from a narrow set of theoretical expectations, I analyzed the creation of markets for ecosystem services by concentrating on those people who are actively trying to get these institutions up and running, and the policies, processes and products they are developing in their efforts.

Following this approach, it is natural to choose the Chesapeake Bay watershed, the Willamette River basin and the Ohio River basin, since these are the regions in which the most prominent and well-resourced attempts at market creation have taken place. These are the three places where the community of market proponents and enthusiasts has exerted most of its efforts to make markets work in ways that meet their lofty hopes and aspirations. Efforts in these regions were intended to lead to trading in multiple types of credits and different kinds of ecosystem services, seeking to fulfill the promise of a more complete accounting of the full breadth of the environmental benefits nature provides to humans. The ideal of multiple credit types also appeals to the goal of many landowners and their advocates, to enable a reliance on multiple streams of revenue. My choice to study actual market-creating practices, as well as the connection between my chosen methods and the studied markets, are relatively uncommon and are described in some detail in the following sections.

Practical methods

The idea of practice as I use it here emerges from interpretative policy analysis, where it has become a central concept that brings together the rules and structures of institutions with the behavior and actions of individuals.[31] In focusing on practices, I do not follow any theory-derived rules or definitions to understand what a market for ecosystem services should be or to analyze why a market is or is not meeting those standards. Instead, I argue that the

---

31 For detailed arguments about why *practice* can be seen as central to the way in which people experience and shape policy, and should therefore be at the center of policy analysis, see: Fischer and Forester (1993) and Hajer and Wagenaar (2003). The older and related idea of *practice* as the place where knowing and doing, or *education and experience*, come together can be found in Dewey (1963). "Practice" is a complex word with a long history and many definitions. Within the field of policy analysis, influenced by the work of French sociologist Pierre Bourdieu (1977; Bourdieu and Wacquant, 1992), it has come to be used over the past 20 years to describe the connections between action, methods and theory, in that order, as opposed to starting from theory, and then moving to methods and action (Corburn 2005, 9).

people working to develop and expand markets for ecosystem services are in fact negotiating their goals and rules as they create them. The phenomenon of the ecosystem service market, in short, gains definition through the actual work of the actors.

The methodology I have used to study markets for ecosystem services can be described as a sort of multi-sited ethnography.[32] I do not use the term "ethnography" to imply that I set out to describe the cultural elements of markets for ecosystem services. Rather, I started with the observation that the practitioners working on the creation of markets for ecosystem services are part of a single, highly connected network of people.

Looking for people who have organized and attended the Ecosystem Markets Conference since its inception can most easily identify the core group of people who have worked on the development of markets for ecosystem services in the United States. Some of them have moved to different organizations since the first workshop, but have continued to work on the creation of markets for ecosystem services. One of the organizers of the first workshop, held in 2008, did so in his capacity as manager of conservation incentives at the American Forest Foundation. He is now a senior associate in the Ecosystems and People program at the World Resources Institute, and continues to be deeply involved in efforts to create markets and in the organization of the conference. Another leading market proponent, initially an environmental marketplace analyst with Clean Water Services who presented on the Tualatin program during the first workshop in 2008, is now the executive director of the Willamette Partnership, the primary organization working to create markets in Oregon. So to study the challenges associated with the creation of markets for ecosystem services, I basically *follow the actors*,[33] in this case the advocates, officials and entrepreneurs working on the creation of specific markets for ecosystem services.

To develop a detailed and accurate understanding of the challenges of creating ESMs, I relied on a combination of interviews, ethnographic accounts and document analysis. I conducted 75 semi-structured interviews with people involved in the creation of markets in the three areas I studied. Besides

32 The history of this approach is connected to the emergence of globalization, and the questions this raises for conventional anthropology. In their chapter "Taking Account of World Historical Political Economy: Knowable Communities in Larger Systems" (1986), anthropologists George Marcus and Michael Fisher contend that this approach is especially needed to make sense of large-scale political and economic processes, which I take the marketization of ecosystems to be. A more detailed and extensive description of this approach was later produced by George Marcus in his *Ethnography Through Thick and Thin* (1998).

33 This technique can be traced to classic anthropological accounts (Malinowski 1922) but has become popularized and immediately relevant to this analysis through its use in science and technology studies (Latour 1987).

interviewing these individuals, I had the opportunity to observe some of them in action. Given the important role played by a relatively limited number of people, I decided to attend and participate in several conferences where the community gathered, such as the Ecosystem Market Conferences in 2011, 2012 and 2014. This allowed me not only to observe the arguments that key thinkers in the field are inclined to make, but also how experiences and ideas about emerging best practices move around from place to place.

Since the ecosystem service markets I studied in detail have the active support and involvement of government agencies, much of the documentation related to these three market-areas is public. This meant I could look at minutes and sometimes transcripts of key meetings. Finally, I visited some of the restoration sites and gained a firsthand sense of the "products" these markets are generating. I did not observe what was happening at these sites with the goal of evaluating restoration efforts from an ecological point of view. Rather, my goal was to understand what environmental restoration looks like in the context of an ESM.

My study methods reflect the fact that this is not the kind of situation in which a traditional comparative case study—looking at similarities and differences between cases—is likely to be very instructive. Simply put, experiences in one effort to create a market are typically relayed rapidly to actors in other places, and some individuals work actively in multiple, geographically dispersed, market creation projects. A straightforward comparison of markets in various regions would render these connections invisible when they are in fact quite central to the ways in which the market creation work has developed over the past decade.

People and markets

The market creation efforts central to this book are the Bay Bank and the Nutrient Trading Program, both in the Chesapeake Bay watershed; the Ohio River Basin Trading Project; and the Willamette Marketplace in Oregon. The Chesapeake Bay watershed covers (parts of) six states, namely, New York, Delaware, Pennsylvania, Maryland, Virginia and West Virginia. The Ohio River flows through or borders six states, namely, Illinois, Indiana, Kentucky, Ohio, Pennsylvania and West Virginia. There are three major reasons for the selection of these market-making projects, despite their differences in scale and scope. The first is that they can be considered, for somewhat different reasons, the most ambitious attempts to bring to fruition the hopes, expectations and enthusiasm regarding ESMs. This selection criterion immediately reveals some of the complexities associated with the ESM idea particularly as it has been pursued in the Chesapeake Bay watershed, and thus warrants a somewhat detailed discussion. The second reason, related to the first, is that these efforts have consistently received a lot of attention at the Ecosystem Markets Conference and in related publications. The third reason is that the protagonists in these efforts have become leaders in the ESM community, with considerable influence over central design elements and important challenges in market making.

The Willamette Marketplace, the Chesapeake Bay Bank and the Ohio River Basin Trading Project are three of the most advanced, if not the only, efforts to create markets in which multiple types of ecosystem service credits can be traded in the United States (Madsen, Carroll and Moore Brands 2010, 21). There are numerous studies on the development of market-like mechanisms revolving around single credit types, like wetland banking, species habitat banking and Water Quality Trading. What makes the Willamette Marketplace, the Chesapeake Bay Bank and the Ohio River Basin Trading Project different is that these are all attempts to incorporate multiple credit types into integrated markets. This characteristic makes them an obvious choice for analysis, since they explicitly seek to achieve the most exacting promises of the market concept.

Incorporating multiple types of ecosystem service credits into a single marketplace is sometimes described as "stacking credits." This metaphor is based on the idea that a single environmental project, like planting trees along a polluted waterway, can produce multiple types of ecosystem services (and therefore credits), like water quality improvements, fish habitat creation and carbon sequestration. Obviously, allowing landowners to sell three types of credits based on a single restoration action could change the economic calculus dramatically. This kind of ecosystem credit stacking is a controversial phenomenon (Cooley and Olander, 2011), and there is very little empirical evidence on the practicalities of stacking credits (Robertson et al., 2014).

The Willamette Marketplace, the Chesapeake Bay Bank and the Ohio River Basin Trading Project were all envisioned as markets in which different types of credits could potentially be "stacked" as shown in Table 1.1. The level to which the implementation of stacking was pursued varies across the regions. For example, the Willamette Partnership successfully developed four pilot restoration projects and calculated and published the various quantities of credits generated on each. The Ohio River Basin Trading Project has conducted in-depth analysis on the idea of combining carbon credits with water quality credits, but in the first 30 pilot projects implemented so far, carbon credits have not been calculated, let alone sold.[34] Despite these varying levels of implementation

---

34 The organization proposing an ESM in the Ohio River Basin, EPRI, conducted a national opinion survey on credit stacking (Fox 2011) and published research on the legal status (Fox, Gardner and Maki 2011; Gardner and Fox 2013) and technical feasibility of combining carbon and water

Table 1.1   Proponents and markets (describing specific credit types where more than one exist)

Location	Willamette River basin	Chesapeake Bay watershed		Ohio River basin
Key Proponent	Willamette Partnership	Pinchot Institute	World Resources Institute	Electric Power Research Institute
Credit Types				
Water Quality	*Temperature*		*Nitrogen & Phosphorous*	*Nitrogen & Phosphorous*
Habitat	*Salmon & Prairie*	*Trout & Forest*		
Wetlands	✓	✓		
Carbon		✓		✓

Source: Kane and Prendergrass 2007; Jones et al. 2010; Willamette Partnership 2009; EPRI 2010.

of the stacking idea, the specific market-making projects are interesting for this analysis since each of them included significant efforts to develop integrated markets where multiple credit types can be bought and sold.

The proponents of the Willamette Marketplace and the Chesapeake Bay Bank are most ambitious when it comes to the number and diversity of credit types they sought to include in their respective ESMs. In fact, several of the Bay Bank's early reports and presentations include references to a desire to incorporate credits for water quality improvements in its marketplace as well. This would mean that a landowner participating in the Bay Bank's proposed habitat credit program could also generate water quality credits. This goal never moved beyond very preliminary analyses conducted by the proponents of the Bay Bank and some online tools developed for the Bay Bank. These online tools potentially allowed for the inclusion of water quality credits as well. The practice of including multiple credit types into a single marketplace proved more complicated, and therefore the inclusion of the World Resources

quality credits (EPRI 2013). Yet, ultimately, in the meetings with policymakers and stakeholders the focus was mostly on water quality credits, and this is reflected in the first round of pilot projects eventually funded. The reductions in greenhouse gas emissions that may have been associated with those pilot projects were not calculated or "credited," nor offered on any kind of market. In 2015, EPRI received a grant from the U.S. Endowment for Forestry and Communities and announced it will pursue another round of pilot restoration projects. For these projects, greenhouse gas emissions reductions will be calculated in addition to the phosphorous and nitrogen reductions related to water quality. See: EPRI's Winter 2015 newsletter, available at: http://wqt.epri.com/pdf/NEWS-LETTER%20Jan%202016.pdf. Retrieved on: 02/01/2016.

Institute's (WRI) work on water quality as a separate effort in this analysis warrants further explanation.

At the time the Pinchot Institute expressed this desire to incorporate water quality credits in its broader attempt to establish an ESM, WRI had already been advocating for and actively working on the creation of water quality credit trading systems, for nitrogen and phosphorous reductions, in the Chesapeake Bay watershed for several years (Faeth 2000). WRI has long focused on developing a combination, or more accurately a *hoped for* combination, of Water Quality Trading programs across multiple states in the Chesapeake region.

The Chesapeake Bay Nutrient Trading Program, as this effort is sometimes called, is based on a traditional water quality market, in which pollution credits can be traded by industrial facilities without any type of ecosystem-based conservation or restoration activity. This ambition alone makes the program a less obvious choice for this study. However, the aspiration of the Bay Bank's proponents to see these efforts integrated makes it impossible to see them as completely unrelated. Furthermore, WRI itself has expanded its efforts to enable the creation of credits by investing in environmentally based projects (so-called non-point source pollution reductions), such as the construction of riparian buffers. This has made its work more closely related to the broader ESM vision and therefore even more relevant to the analysis in this book. In addition, some advocates of WRI's project have also envisioned it as a multi-credit market and have taken steps in that direction. Ultimately, the Chesapeake Bay Bank and the Chesapeake Bay Nutrient Trading Program are listed and described somewhat separately in this book, given the distinct trajectories these two projects have gone through. But both projects reveal some of the profound challenges facing these efforts, despite their different starting points and results.

The second reason to select these four market-making efforts is their central role in national discussions about ecosystem services and the creation of markets. The Willamette Marketplace, the Bay Bank and the Ohio River Basin Trading Project have been at the center of sessions in each of the Ecosystem Market Conferences since the first one in 2008, and the Chesapeake Bay Nutrient Trading Program has featured prominently in the past few years. Two of the most important organizations in this community, the USDA's Office for Environmental Markets and the World Resources Institute, invested significant resources in the creation of advanced measurement systems and tools for the Chesapeake Bay Nutrient Trading Program, making it a testing ground of sorts.

The third reason for selecting the Willamette Marketplace, the Ohio River Basin Trading Project, the Bay Bank and the Chesapeake Bay Nutrient Trading Program is that the organizations behind these efforts have become an identifiable network of thought leaders in the field. Many of the people advocating for and working on the development of ESMs have long been

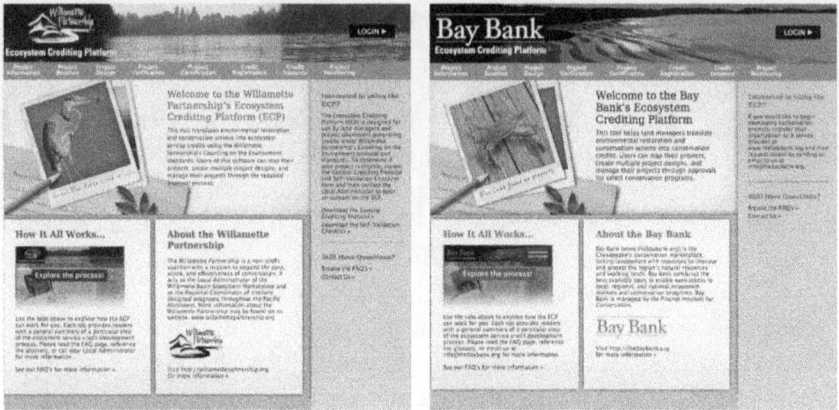

Figure 1.1 Website crediting platforms for the Willamette Marketplace and the Chesapeake Bay Bank. They were developed by a single software development company, The Other Firm, LLC, and form a visual example of the close collaboration between market-creation efforts.

closely connected, personally and professionally. One example of this history of close collaboration is the single Web-based credit registration platform (see Figure 1.1)[35] that the Willamette Marketplace and the Chesapeake Bay Bank share. Another example is the fact that prominent staff members have moved from one organization to the next while continuing to work on ESMs, like a former senior associate at WRI moving to USDA's Office of Environmental Markets or one of the Bay Bank's early advocates now with EPRI, working on the Ohio River Basin Trading Project.

Over the past years, this dense network of market proponents has emerged as something of a brain trust regarding ESMs in the United States. The creator of the Oregon-based market, the Willamette Partnership, has become a highly visible organization, theorizing "best practices" through the publication of reports and presentations. In recent years, the USDA Office of Environmental Markets has contracted the Partnership to author reports on measurement systems for biodiversity markets[36] and Water Quality Trading.[37] That second

35 See: http://willamettepartnership.ecosystemcredits.org and http://baybank.ecosystemcredits.org. Retrieved on: 12/14/2013. The Willamette Partnership website continued to be accessible as of 05/15/2015, but the Bay Bank website is no longer available, probably as a result of limited interest.
36 This report, titled "Measuring Up: Synchronizing Biodiversity Measurement Systems for Markets and Other Incentive Programs," can be found at: http://willamettepartnership.org/wp-content/uploads/2015/04/Measuring-Up-w-appendices-final.pdf. Retrieved on: 05/14/2015.
37 This report, titled "In It Together: A How to Reference Guide for Building Point-non-Point Water Quality Trading Programs," can be found at: http://willamettepartnership.org/publications/. Last accessed on 08/04/2013.

report was coauthored by the Pinchot Institute, the central developer of the Bay Bank, and the World Resources Institute, the organization behind some key elements of the Bay's Nutrient Trading Program. WRI has long been a leader in the realm of ecosystem services, and it developed a widely used set of definitions of specific services.[38] The Electric Power Research Institute (EPRI), the driving force behind the Ohio River Basin Trading Project, has published an extensive series of reports and presentations[39] on its own efforts, and also on existing markets and market-like institutions, becoming an important hub for information on ecosystem service markets in the United States.

Together with a number of other market proponents, the Willamette Partnership, WRI and EPRI have formed the National Network on Water Quality Trading,[40] thereby more formally connecting their efforts to create market-based regimes directly related to ecosystem services. The goals of the network are to develop shared principles, define best practices and capture the diversity of viewpoints around Water Quality Trading in particular, a practice central to the efforts to create more integrated ESMs. This shared emphasis is reflected in the analysis in this book, since in all three regions the market proponents spent much of their time and resources on establishing or expanding Water Quality Trading regimes. But this is not a book about water quality alone, as these efforts were not exclusively about water quality. The broader level of collaboration, information exchange and leadership between the most important organizations trying to create ESMs in the United States highlights the need for an integrated analysis of these institutions, as opposed to a more traditional comparison or narrow focus on a single type of credit.

The Ohio River, the Chesapeake Bay and the Willamette River

The efforts to develop markets for ecosystem services in the Ohio River basin, the Chesapeake Bay watershed and the Willamette River basin are related to long histories of both environmental issues and relevant organizations in those places.

The Ohio River is the largest tributary of the Mississippi River by volume and is nearly 1,000 miles long. Its drainage basin covers 189,422 square miles and is home to more than 25 million people.[41] Twenty dams are on the

---

38 See: http://www.wri.org/project/mainstreaming-ecosystem-services/about. Last accessed on 08/02/2013.

39 See: http://wqt.epri.com/reference-shelf.html. Last accessed on 05/14/2015.

40 See:http://stormwater.wef.org/wp-content/uploads/2014/02/NationalNetworkon WQTOverview_2014-01-10.pdf. Retrieved on: 05/14/2015.

41 Source: Ohio River Foundation Fact Sheet, retrieved from http://www.ohioriverfdn.org/about_the_river/documents/ohioriverfactsversion2.pdf on 05/15/2015.

Ohio River, and major urban centers are located on its banks, like Cincinnati, Ohio, Louisville, Kentucky, and Pittsburgh, Pennsylvania. Based on the federal government's Toxics Release Inventory, the environmental group Penn Environment declared the Ohio River the most polluted in the United States[42] and emphasized the critical role of nitrates from urban and agricultural runoff. Since the Ohio River flows into the Mississippi River, in Cairo, Illinois, this pollution contributes to the Gulf of Mexico's "dead zone." This term refers to fish in the Gulf dying as a result of low dissolved oxygen levels in the water. These levels are low because nutrients like nitrogen foster previously unseen levels of algae growth.

Formal efforts to control and abate the Ohio River's pollution date back to at least 1948, when the Commonwealth of Virginia became the last state to approve the Ohio River Valley Water Sanitation Compact. Seven other states and the U.S. Congress had already approved it between 1939 and 1947. This interstate agreement resulted in the creation of the Ohio River Valley Water Sanitation Commission (ORSANCO),[43] which continues to manage projects and programs to improve water quality in the Ohio River and its tributaries.

While ORSANCO has been involved in the effort to develop a market for ecosystem services in the Ohio River basin, the leading organization was the Electric Power Research Institute. The membership of this research institute is primarily made up of electric utilities. EPRI works on a range of ecosystem service–based approaches, but in the Ohio River Basin the organization focused on water quality and carbon dioxide. While the latter is perhaps more obvious, power plants increasingly discharge nitrogen into waterways as a result of catalytic reduction of nitrogen oxides (NOx). In short, as power generators are reducing their nitrogen exhaust from smokestacks, that nitrogen ends up in water discharge from power plants. In 2006, EPRI issued a white paper (EPRI 2006) summarizing this issue and indicating that power plants could be facing increasingly stringent water quality permit requirements given these discharges. That report also suggested that Water Quality Trading might provide a cost-effective way of meeting such legal obligations. One year later, EPRI issued a second report announcing that the organization would develop a "demonstration project," meaning a "water quality trading program involving an electric power company" (EPRI 2007a, 8–1). Implementation of this demonstration project, for which the Ohio River Basin was chosen as the most appropriate location, started in 2008.

---

42 For the report, see: http://pennenvironment.org/sites/environment/files/reports/Wasting%20 Our%20Waterways%20PA%20Print.pdf. Retrieved on: 05/14/2015.

43 For more information on ORSANCO, and the compact itself, see: http://www.orsanco.org/ orsanco-compact. Retrieved on: 5/14/2015.

The Chesapeake Bay watershed is formed by dozens of rivers and streams flowing into the Chesapeake Bay, stretching from southern New York to Virginia. As the United States' largest estuary, it covers more than 60,000 square miles. The Bay is more than 200 miles long. Large urban centers, including Annapolis, Maryland, and Norfolk, Virginia, are within the drainage area that flows into the Bay. The agricultural activities within the watershed have caused an increase in the level of phosphorous and nitrogen discharged into Chesapeake Bay, resulting in the decline of marine life similar to that in the Gulf of Mexico. Combined with pollution from urban and industrial areas and overfishing, this decline in water quality has adversely affected the Chesapeake Bay ecosystem. Algae blooms are one detrimental consequence of these developments. The tourism and fishing industries, along with important economic activities around the Bay, have suffered as a result.

Organized efforts to protect and restore the Chesapeake Bay watershed date back to at least 1967, with the creation of the Chesapeake Bay Foundation. From the Alliance for the Chesapeake Bay to the Chesapeake Bay Commission, a variety of new entities and forums have been created to improve the state of the Bay. In 1983 the Chesapeake Bay Agreement created the Chesapeake Executive Council as the primary policy-making authority. The first widely publicized call for the creation of markets to reduce pollution in the Bay was a report from the World Resources Institute (Faeth 2000). The organization has been closely involved in the development of trading programs in West Virginia, Pennsylvania and Maryland, primarily through the creation of different versions of Nutrient Net, a tool that facilitates the development of credits.

The Chesapeake Executive Council called for additional action in its 2007 Forest Conservation Directive.[44] This resulted in the Chesapeake Bay Bank, a marketplace that allows landowners to realize financial benefits for the provision of specific ecosystem services. In addition, an independent verification system was created, to make sure that the services provided as a result of landowner actions are credible and measurable. These efforts to create markets for ecosystem services in such a large watershed are among the most ambitious in the United States. They have received significant political attention, in part as a result of its location encompassing Washington, DC.

The Willamette River is a 187-mile long tributary of the Columbia River in the northwest of the United States. Parts of the river basin are densely populated, including urban areas like Portland and Corvallis, but much of the watershed is fertile agricultural land. Channelization and dredging for navigation

---

44 See: http://www.chesapeakebay.net/content/publications/cbp_27761.pdf. Retrieved on: 05/14/2015.

purposes, dam construction for electricity generation and water use for irrigation of agricultural crops are only a few of the interventions that have significantly altered the Willamette River over the past 100 years. The salmonids and resident fish species have been adversely affected as a result of these interventions and have subsequently been listed under the Endangered Species Act.

After the listing, Governor John Kitzhaber created the Willamette Basin Taskforce in 1996, followed by the Willamette Restoration Initiative in 1997. This organization was made up of representatives from federal and state agencies, business leaders, environmental advocates and scientific experts. The Willamette Restoration Initiative announced its comprehensive strategic restoration plan in 2001. In an attempt to create an independent organizational base for the group's work, the Restoration Initiative filed for 501(c) (3) not-for-profit status and changed its name to the Willamette Partnership. The restoration of the Willamette River basin is a complex process, given its ecology, the multiplicity of water uses and the range of interests at stake. Methods to optimize restoration efforts in the basin have been pursued by the Partnership, using the ecosystem services framework to create a series of markets for specific ecosystem services.

1.d Chapter Overview

The three central challenges for the creation of markets for ecosystem services encountered in the Tualatin case are the lack of attention to place in the basic logic of these markets, the tension between creating precise, usable and acceptable measurement systems and the problem of getting the right people to agree to and participate in the market. These challenges provide the core of the answer to my research question as well as the structure of this book. The second chapter concerns the challenges of dealing with place in the creation of markets for ecosystem services. Here, I focus on the expansion of the shade credit trading program in Oregon and the Water Quality Trading program in the Chesapeake Bay, especially in Pennsylvania. The third chapter analyzes the challenges associated with creating measurement systems that can be used in markets for ecosystem services. It looks in detail at the measurement systems for wetlands and prairie in the Willamette Marketplace and those used to enable the creation of ecosystem service credits in the Ohio River Basin Trading Project. The fourth chapter focuses on the problems of finding and engaging the appropriate stakeholders in the development and expansion of these markets. The attempts in relation to the Willamette Marketplace and the Ohio River Basin Trading Project are central to this chapter. Finally, in the conclusion, I summarize what can be learned from the analysis of these challenges and I describe their implications for the practice of creating ESMs.

Chapter 2

CREATING PLACES FOR MARKETS

To achieve the promises associated with ecosystem service markets (ESMs), such as efficiency or an overall reduction in the cost of environmental improvement, it is necessary to displace specific environmental impacts. More precisely, ESMs are based on the idea that it is possible, and even beneficial, to locate compensatory actions away from the point of impact, for example, by planting trees upstream of the discharge of warm water from a treatment plant. This means that operating at a large geographical scale is often one of the basic goals of ecosystem service market creation, since this brings down transaction costs and guarantees a sufficiently competitive "supply" of credits. For example, if a water treatment utility is allowed to purchase "shade credits" from a large area upstream of its point of discharge, the prices of those credits are likely to be lower than when only a few miles of stream are eligible. If more landowners can generate revenue from the sale of those credits, the price of a credit is likely to be lower, as the logic of supply and demand dictates.

The areas, or locations, from where ecosystem service credits can be bought and sold is a central challenge in creating markets for ecosystem services, primarily because of the importance many people attach to *place*. People care deeply and for a variety of reasons about specific locations. So, when markets for ecosystem services treat places as essentially the same, and allow exchanges to happen between them as if all places are fungible, people get upset. For example, a market might allow the discharge of warm water into a salmon-carrying stream to be compensated by planting trees that cast a cooling shade over that river upstream of the water treatment plant. In purely ecological terms, such compensation is noncontroversial. The trade-off benefits the environment; average water temperature is reduced along the entire length of the river. But for people living right next to a water treatment facility or adjacent to a tree-planting project, the impacts will differ dramatically. Such differences can lead to strong opposition, even if a market exchange produces net benefits.

The development of markets for ecosystem services in Oregon, the Ohio River basin and the Chesapeake Bay states has shown how hard it is to deal with the social and cultural meanings of place when creating markets. In

Oregon, this resistance was evident in the lack of support for an ambitious attempt to expand the Tualatin shade credit trading program. The new effort was designed to enable trading at a larger scale and to cover multiple kinds of ecosystem services. But despite many negotiations and much hard work, two key actors decided not to participate in the market: the City of Portland, widely considered one of the "greenest" cities in the United States, and the National Marine Fisheries Service (NMFS), the agency tasked with the protection of endangered salmon species. Both declined to pursue credit trading in the Willamette River basin. Their reasons were related, in part, to concerns about place: Portland preferred to invest in environmental restoration within its city boundaries and NMFS argued that some unique locations are simply critical to the survival of an endangered population of fish.

In the Chesapeake Bay watershed, Environmental Protection Agency (EPA) officials and environmental advocates used arguments about place to criticize the first state to implement an active trading program. They were upset because the market in Pennsylvania consisted of moving chicken manure from polluted areas to other watersheds. Ecologically, the move could be easily justified. Soils in the reception area could be shown to lack the nutrients present in chicken manure. However, the very thought of moving around chicken manure, coded as a pollutant in many people's minds, as a credible environmental solution aroused significant resistance to Pennsylvania's market.

The developers of markets for ecosystem services are well aware of this tension between protecting place and making markets, and they have tried to deal with it primarily by creating rules, sometimes called protocols, about where credit development can happen, which practices can be used to generate credits, and where negative impacts should be avoided completely. Rulemaking has proved complicated, though. It requires incorporating the meaning of specific places—and more generally people's connections to the places where they live, work and play—into the design and operation of markets. All of this runs counter to a central promise of markets: they are supposed to be a policy tool requiring relatively few rules as compared to more traditional "command-and-control" policies and plans.

2.a Places versus Markets

Before turning to the analysis of the resistance to markets in Oregon and Pennsylvania, I want to provide a more detailed examination of the idea of place on one hand and the role of markets in environmental policy in the United States on the other. Markets for ecosystem services are designed to attach value to particular places. To function, markets in effect create *displacements*, by which I mean that the markets move specific elements of one place to another. This can become controversial in different ways. The first, most obvious tension grows out of the observation that there are certain unique or even transcendental places that people believe simply ought to be protected against any kind of degradation. National parks are the most immediate example of this type of concern, as are unique habitats and sacred spaces. A second, related source of tension is that many public parks and urban green spaces have specific meaning for residents and users and are deeply valued where they are. While such places are not quite as obviously deserving of preservation as Yosemite Valley or the Grand Canyon, the value these places have to local communities does not very easily lend itself to translation in an ESM. Finally, even places that are already part of an explicitly economic logic, such as agricultural lands and abandoned mines, do not always lend themselves to incorporation into a market for ecosystem services. Removing a polluting source from one place to another can affect the meaning of both places in unexpected ways even across very short distances.[1]

This emphasis on place is related to, but not exactly the same as, another argument about the challenges ESMs face, namely, about their scale. Several scholars have pointed out that the desire to create these institutions on a large spatial scale creates problems. For example, large-scale markets hinder the ability to develop strong social ties and interpersonal trust between credit buyers and sellers (Moore 2014). Other research points out that compensation activities might cluster in small, concentrated areas far away from the sites of impact, resulting in an overall shift in landscape quality (Robertson and Hayden 2008; Ruhl and Salzman 2006). While these arguments are in line with my finding that the logic of displacement causes resistance to ESMs, it is not exactly the same. A mere reduction in the scale of ESMs, for example, by restricting trading to a small watershed or even within a county, is unlikely to alleviate the concerns over what makes a particular place special or even just useful.

A sense of place

To explain the strong objections to trading ecosystem credits across geographical locations, even over short distances, it is instructive to start with the movement to protect landscapes in the United States. Some of the most significant and enduring efforts to preserve landscapes took place in the Potomac River basin in Pennsylvania and Maryland. This river connects much of the Chesapeake Bay watershed. Its tributaries in Pennsylvania, West Virginia and Virginia flow into its main stem. Early landscape preservation efforts in the

---

1 This argument is widely associated with Mary Douglas's work on pollution, but can also be found in Michel Callon's analysis of a series of attempts to grow scallops in St. Brieuc Bay (Callon 1986).

Potomac River basin started during the American Civil War and were a direct result of it. Two years after the Battle of Antietam, named after the eponymous creek that runs through the center of the battlefield and flows into the Potomac, a plan was introduced to create an official national cemetery in that location.[2]

The successful attempt to preserve this unique place, and comparable efforts elsewhere, were in some ways a precursor to the creation of much larger national parks such as Yellowstone and Yosemite. At Antietam, Gettysburg and other battlefields, the federal government assumed responsibility for the preservation of significant tracts of land in unprecedented ways (Faust 2008, 99–100; Mackintosh 1987). The protection of Antietam Creek and its surrounding areas, where the bloodiest single day of the Civil War took place, reveals an important element of what planning scholars refer to as "socio-spatial relations" (Graham and Healey 1999). This concept refers to the observation that people attach profound importance to the way in which particular locations are assembled. In the case of Civil War memorial sites, historical significance defines the unique meaning of those locations.

Decades after the Civil War, the need for a national imperative to protect particular places (now for their environmental significance) was central to John Muir's writings on Hetch Hetchy and the Yosemite Valley (1912). More recent examples of ethical arguments for the protection of more mundane places can be found in Aldo Leopold's "Land Ethic" (1949) and the school of thought known as "deep ecology" (Devall and Sessions 1985; Naess 1990). Leopold, Naess and others have long argued that certain locations are inherently worth protecting; therefore, disturbances of all kinds should be avoided in these locations. This ethical point of view has well-developed theoretical underpinnings that go beyond mere environmental activism. To those who believe in the uniqueness of places, whether for historical, ecological or ethical reasons, swapping environmental goods and bads is problematic. The particular *socio-spatial* configuration of a place, be it a battlefield or an aesthetically pleasing valley, carries moral significance that, in the minds of many, should not be altered.

Place is also important in environmental thinking because proximity to nature matters, even (or especially) when speaking about small parks in urban settings. Small green spaces are just as important to some people as well-known protected areas of national significance, such as Antietam or Yellowstone. A classic philosophical argument about being able to retreat to natural areas with relatively little human activity goes back to Thoreau's *Walden* (2008)

2 For a brief description of the history of the national cemetery at Antietam, see http://www.nps. gov/anti/historyculture/antietam-national-cemetery.htm. Last accessed on 05/18/2015.

[1854]. In his famous account, the ability to spend time in nature is at the heart of a quest to lead a simple, deliberate life. More recently, proximity to parks and gardens, and their presence in urban areas, has been central to writing about good city planning (Lynch 1981; Spirn 1984). Advocates of urban parks and the protection of city landscapes argue that these are important to the well-being of communities and to public health (Frumkin 2003). This line of reasoning emphasizes the need for green spaces close to where people live, because it makes them more social as they meet each other in public parks and more healthy as they walk to, from and through these spaces. This reasoning highlights a potential problem regarding markets for ecosystem services. Trading the degradation of urban green spaces for the restoration of rural landscapes, while potentially effective from a trading perspective, may have a disproportionate adverse impact on populations in cities.

A third source of questions about the feasibility and desirability of markets for ecosystem services in relation to place is linked to the dynamics of change. Anthropologist Mary Douglas (1966) first suggested that something becomes *pollution* once it is viewed as *matter out of place*. This view of pollution emphasizes the importance of local values in determining environmental goods and bads. Douglas's approach to understanding pollution explains why simply moving matter to a *different* place is not necessarily a cure. From this point of view, warm water cannot be considered environmentally good or bad without taking into account its location and the meaning it holds in that location. Warm water is not generally considered a pollutant. It only becomes one once it enters a stream used by fish that require cold water for survival, as is the case in the Tualatin River and much of the Pacific Northwest of the United States. This paradox is recognized in the law: water temperature becomes polluting once it enters a stream considered impaired given its designated use, as mandated under Section 303 of the Clean Water Act, and then only if it surpasses the Total Maximum Daily Load (TMDL) allocation.

The question of displacement—in each of the three ways I describe—is at the heart of many environmental controversies. It helps to explain enduring conflicts in environmental policy and rulemaking. For example, with regard to air quality regulation:

> attempts to resolve one environmental problem (for example, by building tall smokestacks to reduce local pollution) often simply exacerbate another kind of problem (for example, long-distance pollution such as acid rain). (Paehlke and Torgerson 2005, 85)

Since markets for ecosystem services rely on this same logic, even if proponents can show efficiency gains and overall effectiveness, it is not surprising that moving matter in and out of place arouses some resistance to the creation and reliance on such markets.

In summary, the importance of place can manifest itself in three ways in the design and implementation of ESMs. The first is when people simply believe a location is unique for historical, ecological or any other reasons and therefore reject possible changes in that place. The second grows out of concern about the benefits of proximity to nature, especially for urban populations. The third is based on the idea that altering a set of socio-spatial relations, by moving matter from one place to another, might create new forms of pollution. These long-standing debates raise practical questions about what the appropriate geographic location is for each credit and debit traded in a market, and which elements should and should not be moved around. All of this runs counter to the logic of matching supply and demand in the most flexible way to efficiently achieve specific environmental goals.

The market in environmental policy

The logic and language of markets has seen increased attention over the past 50 years. The proponents of ESMs did not start their efforts from scratch. The Willamette Partnership, the Electric Power Research Institute (EPRI), the Pinchot Institute and the World Resources Institute (WRI) all sought to integrate or significantly expand existing environmental trading regimes and market-like mechanisms in their projects. In order to understand how their efforts to create ESMs built on these existing markets, it is instructive to provide a brief overview of the history of the most important environmental trading regimes in the United States.

An important starting point for ESMs was John Dales's work on regulatory pollution markets in 1968. But since the late 1960s, the ideas and ideals associated with the use of markets have both deeply penetrated and transcended the realm of environmental regulation (Sandel 2012). More recent theoretical approaches to policy implementation, like the *new governance* (Salamon and Elliott 2002; Salamon and Lund 1988) and *new public management* (Dunleavy and Hood 1994; Hood 1986; 1991), call for a wide range of market and market-like elements. In these lines of thinking, markets are still contrasted with command-and-control or tax-based approaches to regulation, but cap-and-trade systems are not the only envisioned alternatives. A more diffuse set of ideas, including the use of private sector management styles, performance controls and output measures, has become part of the lexicon associated with markets (Cashore, Auld and Newsom 2004; Wu and Babcock 1996). Outside of academia, the rise of the "New Right" of Ronald Reagan and Margaret Thatcher is usually seen as the moment when the supposed superiority of markets and the private sector over state regulation gained widespread political popularity (Conniff 2009; Hood 1991; Layzer 2012).

There have been many efforts to incorporate market and private sector values into environmental policy implementation. The proponents of ESMs in the Willamette River basin, the Chesapeake Bay watershed and the Ohio River basin have been most interested in existing market-like mechanisms under the Clean Water Act (CWA) and the Endangered Species Act (ESA). Three separate mechanisms have emerged under the CWA and ESA, and each one has its own complex legal, scientific and institutional history. Water Quality Trading and wetland banking, as the two mechanisms under the Clean Water Act (under Sections 303d and 404, respectively) are generally called, started in the early 1980s. ESM proponents in the Willamette River basin and the Chesapeake Bay watershed have also sought to incorporate a third trading mechanism, namely *species habitat banking* or *conservation banking*. The U.S. Fish and Wildlife Service leads the development of this market-like mechanism, under the Endangered Species Act (Sections 7 and 10). The history of these three market mechanisms highlights the complex and often contested regulatory and institutional terrain that the ESM proponents entered.

The first pilot project with Water Quality Trading started in Wisconsin in 1981 (McGinnis 2001). After that early effort, dozens of similar systems have been developed, with varying levels of success as described in Chapter 1. A central element of Water Quality Trading and more generally the implementation of Section 303(d) of the Clean Water Act is setting the water quality criteria for each water body; the so-called Total Maximum Daily Loads (or TMDLs). Individual states are responsible for setting TMDLs, meaning the assessment of how many pollutants a factory or a power plant can discharge while still assuring the *biological, chemical and physical integrity of the nation's waters*, the Clean Water Act's statutory goal. For trading to work, the TMDL for a specific water body needs to be clear, precise and consistently enforced through the permitting process. Based on the TMDL for a water body, specific limits for each entity discharging water into that stream or river need to be set. Otherwise, water dischargers have little direct incentive to look for ways to reduce the levels of those pollutants, either by upgrading their water treatment procedures or by buying credits.

Determining these water quality criteria and translating them into precise limits for each wastewater discharge permit in a watershed is a scientifically complex, politically charged and labor-intensive process (Houck 2002, 194–195), and its completion and efficacy vary dramatically from state to state. This has long been a source of concern for close observers of the Clean Water Act. For example, one expert expressed that it was "crunch time for Water

Quality Trading" even as the enthusiasm for ESMs with multiple credit types was emerging (King 2005).

Wetland banking, the second market-like mechanism connected to the Clean Water Act, can also be traced back to the early 1980s but has become much more widespread than Water Quality Trading in the intervening years. The basic idea behind wetland mitigation banking is that damage to a wetland as a result of development, like the construction of a highway or shopping mall, can be compensated for. Accepted forms of compensation are the establishment of a new wetland, the restoration of a wetland that existed previously, the enhancement of a current wetland or the conservation of one. In 1981, the U.S. Fish and Wildlife Service published its mitigation policy in the Federal Register[3] and a few years later the Tenneco LaTerre Corporation created the nation's first wetland bank (Dennison 1997, 138–140), in Terrebonne Parish, Louisiana. Initially, a limited number of state transportation agencies and large corporations created wetland banks to compensate for their own construction activities (Hough and Robertson 2008). Over the past two decades, more entrepreneurial organizations have proactively restored wetlands in order to sell "credits" to multiple other organizations in need of compensation (Robertson 2009).

In contrast to the TMDL program and Water Quality Trading, the responsible agency for Section 404 and the wetland-banking program is a federal agency, the U.S. Army Corps of Engineers, with input from the Environmental Protection Agency. This means that demand for wetland "credits" is directly related to the requirements those agencies set forth. Several elements of the permitting process for discharging dredged or fill materials into wetlands have been contentious since the inception of the mechanism. These elements include the exact specification of the location, quantity and quality of wetlands that need to be established, restored, enhanced or conserved (BenDor and Brozović 2007; King and Herbert 1997; Robertson and Hayden 2008). These contestations have resulted in a decades-long series of lawsuits and rule-making efforts (Lewis 2001).

Despite its contentious history, wetland banking has become an established practice. Between 1995 and 2014, the U.S. Army Corps of Engineers approved more than 1,400 mitigation banks (Institute for Water Resources 2015). More than 1,000 of those mitigation banks are commercial, meaning that the credits from the banks can be sold to different developers who need to compensate for their impact(s) on wetlands. The overwhelming majority of these commercial banks are owned and operated by private entities, generally

---

3 See: U.S. Fish and Wildlife Service (1981). U.S. Fish and Wildlife Service mitigation policy; notice of final policy. Fed Register 46:7644–7663.

for-profit organizations. These statistics indicate that wetland mitigation banking is a vibrant, market-like mechanism, especially when compared to the more tenuous status of many of the 70 or so Water Quality Trading regimes.

The third and final existing market mechanism under American environmental law of interest to several of the ESM proponents is species habitat banking. This practice is similar to and very much inspired by wetland banking in that development of sensitive areas can be compensated for by restoring, enhancing or conserving land elsewhere. The first effort to develop a species habitat bank was to protect the California gnatcatcher, proposed for listing under the Endangered Species Act in 1993. An area of 263 acres in the vicinity of San Diego, known as Carlsbad Highlands, was preserved. This produced 180 Coastal California gnatcatcher credits, of which 83 were eventually sold to various organizations, including California's Department of Transportation, to mitigate the impact of developing projects in known gnatcatcher habitat elsewhere in Southern California.[4]

That same year, the California EPA and the California Resources Agency jointly issued the "Official Policy on Conservation Banks" (Wheeler and Strock 1995) to clarify and formalize the general rules under which this new type of market would operate. After a range of different banks was established in California, using somewhat different legal and scientific approaches (Mann and Absher 2014), national rules were established in 2003 with the publication of the U.S. Fish and Wildlife Service's "Guidance for the Establishment, Use and Operation of Conservation Banks." Despite the existence of national policy on species habitat banking, this practice continues to be largely focused in California. By 2011, a total of 132 conservation banks had been established, for species ranging from the Florida panther to the Carolina heelsplitter mussel (Madsen, Kandy and Bennett 2011, 6–7). The overwhelming majority of these banks, 89 to be exact, was located in California, making this an exceedingly rare practice outside of that state.

These three existing market mechanisms are complex and somewhat controversial in their own right. Their existence provided the organizations seeking to develop ESMs for multiple credit types in the Willamette River basin, the Ohio River basin and the Chesapeake Bay watershed specific avenues and a general approach to build on when developing integrated markets. Yet in their statements about the goals of creating ESMs for multiple kinds of ecosystem service credits these advocates generally used the language of markets and market-like elements in broad terms, and only vaguely referenced the

4 More information on this species habitat bank and others in the United States is available at us.speciesbanking.com. Last accessed on 03/21/2016.

detailed experiences with Water Quality Trading, wetland banking and species habitat banking.

Efficiency, innovation and private participation

The goals and promises associated with the creation of markets for ecosystem services have been wide-ranging. They sometimes differ dramatically, though, in the eyes of specific groups or organizations involved in creating new ecosystem service markets. Whenever groups like the Willamette Partnership, the Electric Power Research Institute (EPRI) and the World Resources Institute (WRI) have tried to create markets for ecosystem services, they have keyed their proposals to reduction of direct government control, increasing efficiency or stimulating innovation—and sometimes to all three. The exact ways in which ESMs can or will achieve these and other goals are not always clearly spelled out. Further complicating matters is that whether these goals are achieved is hard to measure (Chichilnisky and Heal 2000). This has led some scholars to argue that too much rhetoric occurs about cost-effectiveness, revenue production and environmental improvement that is largely disconnected from practice and the experience with market mechanisms under the Clean Water and Endangered Species Acts (Stephenson and Shabman 2011).

The official statements of main proponents of the creation of markets in Oregon, the Ohio River basin and the Chesapeake Bay states reflect this lack of clarity about how to connect the theoretical goals of market creation to their actual functioning and the long history of market making under the Clean Water and Endangered Species Acts. For example, the Willamette Partnership, the organization most actively working on the creation and expansion of ESMs in Oregon, describes what "ecosystem services markets do" as follows: "They [ESMs] marry the economy and the environment, creating new business opportunities while increasing the pace, scope, and effectiveness of conservation and restoration."[5] Exactly what this marriage looks like, though, or whether there is an equal division of household chores between the spouses, is not clear. Increased effectiveness does imply that more conservation and restoration can be achieved with the same amount of money or less.

While not explicitly promising an overall reduction in regulatory burdens on water treatment plants or developers, the Partnership does present the ESM as an alternative way of meeting regulatory standards: "Environmental regulations set standards to protect natural resources. Industries, businesses, developers, and individuals who change the land or water must either meet these regulatory

---

5 For this and other goals of ecosystem services markets described by the Willamette Partnership, see: http://willamettepartnership.org/market-tools-rules/ Last accessed on 05/18/2015.

standards or compensate for the impacts they cannot avoid."[6] Again, it is not immediately clear how the relationship between the theoretical market and its operational rules is supposed to work, especially given the complex histories of creating market mechanisms to implement those regulatory requirements.

Similarly, an important group of officials from the various states involved in the effort to create nutrient trading programs in the Chesapeake Bay has written:

> Environmental markets show promise for protecting and restoring clean air and water, wetlands, healthy wildlife, and a suite of other environmental benefits while encouraging innovation and private investment, improving accountability, reducing costs, and expanding conservation opportunities for landowners.[7]

Efficiency, innovation and private participation are explicitly mentioned here as well. These and other similarly broad claims about the reduced need for government intervention, increased efficiency and potential innovation associated with the use of markets have been translated into ideas for how to design markets for ecosystem services in Oregon, the Ohio River basin and the Chesapeake Bay states.

The tension between these broad goals and the importance of place in the history of environmental thinking and advocacy might not be immediately apparent. But the implementation of market goals has, in all three regions, emphasized the need to create significant flexibility in determining where exactly ecosystem service credits are generated and debits are incurred. The focus has been on creating markets at a large enough geographic scale to encourage new ways of developing credits by moving specific elements of nature from place to place. These design guidelines follow theoretical market values and logic, but are at odds with the ideals of place that have emerged as part of environmental theory over the past 150 years. The following sections describe how these tensions played out in specific ways in Oregon and the Chesapeake Bay states (Figure 2.1).

2.b Ecological Outsourcing in Oregon

To understand the importance of place in efforts to develop real markets for ecosystem services, it is instructive to return to Oregon, in 2004, when the

---

6 See: http://willamettepartnership.org/market-tools-rules/ Last accessed on 05/18/2015.
7 The Chesapeake Bay Environmental Markets Team is an intergovernmental collaboration of 12 federal agencies, established as a result of Executive Order 13508. For the full text of the Charter, see: http://www.usda.gov/oce/environmental_markets/files/FY%2010–11%20CB%20EMT%20 Charter_final.pdf Last accessed on: 05/19/2015.

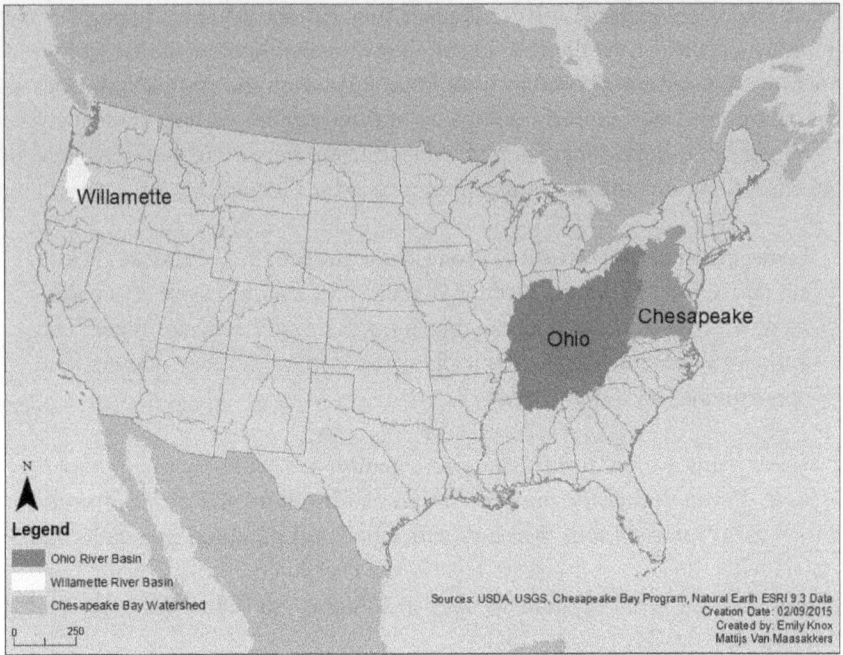

Figure 2.1 The three regions in the United States where the efforts to create markets for ecosystem services have taken place.

first shade market was created. In that year, Clean Water Services received its permit allowing the development of shade credits, thereby avoiding the need to build an expensive "chiller," as described in the previous chapter. In 2005, the Willamette Partnership, the organization that would later become a central member in the national community of proponents of ecosystem service markets, received a "Targeted Watershed Grant" from the EPA.[8]

The primary goal of the work to be funded with this $779,000 grant[9] was to create a program based on the success of the Clean Water Services shade market, but now in the much larger Willamette River basin.[10] The Tualatin River is 83 miles long, and its drainage area covers 712 square miles. In comparison, the Willamette River is 187 miles long, and its basin covers more than 11,500 square miles.[11] The scale of the Willamette River basin, and the desire to reduce

8 This "National Pollutant Discharge Elimination System" permit is described in Chapter 1 as central to the "Tualatin story."

9 See the EPA Factsheet on the Targeted Watershed Grant for the Willamette River Basin. EPA840-F-07-001H. Available at: http://water.epa.gov/grants_funding/twg/upload/2007_04_04_watershed_initiative_2005_willamette_river.pdf Last accessed on: 05/19/2015 Last accessed on 05/22/2015.

10 Ibid.

11 For detailed spatial and environmental information on the watersheds in Oregon, see: http://oregonexplorer.info Last accessed on 05/20/2015.

transaction costs, led the Willamette Partnership to pursue an ambitious integrated marketplace, including multiple types of ecosystem services and credits. After much effort was expended to convince the relevant organizations and officials, two important representatives nevertheless decided not to support implementation of this marketplace, citing considerations related to the idea of place.

From the Tualatin to a marketplace

Initially, much of the work under the EPA's Targeted Watershed Grant focused on the expansion of the temperature credit program. To deal with the issue of scale, and to prioritize the need to ensure significant supply and achieve efficiency, the Willamette Partnership proposed two ways in which concerns over place would be dealt with, namely the designation of *service areas* and *priority areas*.

Service areas set geographic boundaries within which trades must be made. They specify the overall scale of the market. For example, a water treatment facility in need of shade credits can only purchase them if the restoration project in question is located within the relevant service area. The Willamette Basin was divided into three "service areas." Each was framed in terms of the water quality standards set by the Oregon Department of Environmental Quality following the requirements in the Clean Water Act. These standards are outlined in the Total Maximum Daily Loads (TMDL). Within each area, the Willamette Partnership emphasized the need to provide as much flexibility as possible regarding where credits should remain in order to "improve ecological effectiveness, administrative efficiency and credibility of marketplace activity" (Primozich 2008, 49).

To encourage implementation in certain places, however, priority areas were defined. These places were considered important for environmental reasons; restoration projects in these areas were considered particularly valuable. To achieve this kind of targeting, the Willamette Partnership and its partners suggested percentage requirements that had to be met. This means that a specific proportion of the credits purchased by a water treatment utility like Clean Water Services would have to come from tree-planting projects in priority areas. These conditions were described in detail in the final report submitted to the EPA (Primozich 2008).

The completion of the EPA grant-project and publication of this document did not, however, result in the establishment of a shade market in the Willamette Basin. Instead, participants tried to create a more integrated *marketplace*, including markets for wetlands, carbon and an endangered species habitat (Primozich 2008, 51; and see Table 2.1).

Integrating these markets would raise new difficulties and, given their contentious legal histories and diverse ecological objectives, integration of

Table 2.1　Federal statutes and regulatory agencies in *Counting on the Environment*

Credit Type	Regulatory Driver	Agency
Prairie Habitat **Salmon Habitat**	*Endangered Species Act, Sections 7 & 10*	*U.S. Fish and Wildlife Service National Oceanic and Atmospheric Administration – Fisheries*
Water Quality— Temperature	*Clean Water Act, Section 303(d)*	*U.S. Environmental Protection Agency*
Wetlands	*Clean Water Act, Section 404*	*U.S. Army Corps of Engineers*

Source: Willamette Partnership, 2009

such trading schemes was a highly complicated proposition. The Willamette Partnership's final report to the EPA explains why it nevertheless sought this expansion in scope:

> Two types of costs predominate in most markets. One is the payment made to sellers to compensate them for providing the good or service. The other is payments to intermediaries or other costs such as time, processing, fees, and appraisals. The latter are called transaction costs. A fundamental requirement for markets to emerge is that there must be gains from trade for the buyer and seller, after adding up the costs to participate. (Primozich 2008, 52)

This emphasis on transaction costs, a central concept in economic theory (Coase 1937; Williamson 1981), highlights the way in which the Willamette Partnership and its supporters followed market-based economic theories to achieve environmental goals. The incorporation of wetland banking, carbon markets and endangered species habitat proved formidably complicated, but the underlying economic argument—that transaction costs should be minimized—was compelling enough to lead them to try.

Before their report was published, the Willamette Partnership had already received the next large federal grant supporting further development of the ecosystem service market in the Willamette River basin. This was a Conservation Innovation Grant (CIG) from the Department of Agriculture's Natural Resources Conservation Service (NRCS), awarded in 2007.[12] This

---

12　For a brief description of the original goals and size of the grant, see the NRCS website: http://www.nrcs.usda.gov/wps/portal/nrcs/detailfull/?ss=16&navtype=SubNavigation&cid=stelprdb1044500&navid=100120290000000&pnavid=100120000000000&position=Not%20Yet%20Determined.Html&ttype=detailfull&pname=2007%20Conservation%20Innovation%20Grants%20Awards%20|%20NRCS Last accessed on 05/19/2015.

new effort was named Counting on the Environment (COTE). This was a stakeholder involvement process led by a professional mediator[13] given the large number of state and federal agencies, municipalities and advocates involved. The diversity of stakeholders reflected the broad scope of COTE. Its goals included incorporating markets for wetlands, carbon and endangered species into a single marketplace. COTE involved several daylong negotiation sessions in 2008 and 2009 with approximately 30 agencies, advocacy groups and municipalities.

These negotiations resulted in the *Pilot General Credit Protocol: Willamette Basin Version 1.0.*[14] This document described the service and priority areas in some detail, not only for shade credits, but also for wetland credits, salmon habitat credits and prairie credits.[15] Instead of percentage-based requirements regarding the purchase of credits from priority areas, the report suggested ratios. The ratio-based approach incentivized the protection as well as the restoration of priority areas. For example, if an environmental impact occurred in a priority area, the developer would have to purchase credits on a 1:2 ratio, meaning two credits would compensate a single "debit." A similar ratio was applied to the development of credits, meaning that a landowner who was restoring in a priority area would effectively receive more credits to sell than a landowner implementing a similar restoration project outside of the priority area. This approach still left ample flexibility about exactly where impacts and restoration projects would occur. The fact that this new marketplace included multiple types of credits based on different laws meant that each credit type had its own service and priority areas and sometimes required different ratios.

---

13 Mediation of complex environmental decision-making processes is a professional field in Oregon. A description of a variety of processes that have been supported in this way, including the Counting on the Environment process can be found at: http://www.oregonconsensus.org Last accessed on 05/22/2015.

14 The protocol as well as the supporting documents and the formal agreement can be found at: http://willamettepartnership.org/market-tools-rules/ Last accessed on 05/19/2019.

15 The marketplace envisioned in this document included several types of ecosystem credits. Therefore it also relied on different laws and regulations that established the need to "offset" the impact of a project, and thus different service and priority areas. These regulations had different implications for the scale at which trades could take place, and which areas were considered most important. For example, the requirements to compensate for the discharge of warm water in much of Oregon as well as the requirement to compensate for the destruction of a wetland almost anywhere in the United States are both based on the Clean Water Act. But for wetlands, restoration projects generally have to be within the so-called *4th field hydrological unit*, which are sub-basins with an average size of 700 square miles, comparable to the Tualatin River watershed. Temperature trading is driven by a different section of the Clean Water Act, and the scale at which trades can potentially take place under this section varies dramatically, from the Tualatin Basin to the Chesapeake Bay watershed. For prairie habitat, the suggested area within which trades could take place was based not on a law, but on the idea of an eco-region, as defined by the Nature Conservancy's Eco-Regional Plan. The Nature Conservancy also publishes an integrated priority area map, which included different priorities for preservation and conservation of natural areas.

The overall logic was consistent across the different credit types, reflecting the prioritization of efficiency and flexibility over place-based concerns, but the inclusion of priority areas and trading ratios reflected a realization that not all places could be treated the same way.

The main organizers of the COTE process were the Willamette Partnership's executive director at the time and a Clean Water Services representative, who now directs the Partnership. These two men worked to formalize the protocol and got agencies and representatives to sign an *agreement in principle* to test and pilot the marketplace. Representatives from the various organizations and agencies that had participated in COTE viewed this as an important symbolic moment in which they would move from negotiation to implementation. It was also, finally, an attempt to move from theory to practice.

Localized resistance

Despite the initial enthusiasm and publicity around the Healthy Streams Initiative in the Tualatin basin, and the extensive participatory efforts generated by the EPA-funded project and the ensuing COTE process, it was difficult to find places where ecosystem services could be created and traded in the Willamette River basin. Two important entities, both of which had been involved throughout the market design processes, eventually declined to participate or sign the *agreement-in-principle*. These were the National Marine Fisheries Service (NMFS) and the City of Portland. To illustrate how place proved a formidable obstacle to market creation, I will review their objections in some detail.

The NMFS representative was the first to make clear that he would not sign the agreement. NMFS is the primary federal agency tasked with developing and implementing recovery plans for salmon populations under the Endangered Species Act. In addition, salmonids are central species to many broader environmental preservation efforts in the Pacific Northwest (Taylor 1999). Indeed, the shade credit market is indirectly driven by concerns over salmon survival and water temperature, so the refusal of the agency primarily responsible for the protection of endangered salmon to participate in the creation of the new marketplace was a major blow to the market's success. When asked why he would not sign, the NMFS representative pointed to the significance of place or, more accurately, of certain unique places:

> In Willamette Valley alone, we have over 107 populations listed of threatened and endangered salmon and steelhead, so we don't really feel that a site with a certain characteristic of water temperature and riparian

conditions is the same throughout the Willamette. (Interview with author, July 26, 2011)

Thus, his argument was that adverse impacts on certain salmon species, given their very specific locations in the valley, are virtually impossible to compensate for in other places. Moreover, if improvements elsewhere would degrade water quality in Portland, that single negative impact would harm every species:

> The fact is that every individual of all of those 107 populations still has to go through downtown Portland and it just doesn't make sense to trade physical habitat improvements that are not necessarily the limiting factors for recovery of those populations in exchange for further degradation of water quality here which is the bottleneck to the entire basin. (Interview with author, July 26, 2011)

Regulations have not been issued to implement the salmon habitat credit. It seems unlikely to become an approved method of compensating impacts on rivers and streams in the Willamette River watershed. This NMFS representative believes that certain places have unique environmental properties and cannot be traded away. This reasoning echoes the arguments of well-known environmentalists going back to John Muir's case about the ecological and moral significance of the Yosemite Valley, though on a more modest scale. A system of priority areas and trading ratios, however sophisticated, is fundamentally inadequate to protect concerns about uniqueness.

The City of Portland, an important player in environmental efforts in the Pacific Northwest, was the second organization whose representative did not sign the agreement-in-principle. Again, concerns related to place were at the heart of the city's opposition. Portland is a highly progressive city when it comes to sustainability (*Popular Science* 2008) and environmental stewardship (Shandas and Messer 2008). Portland's Bureau of Environmental Services, the agency responsible for water quality and storm water management, was directly involved in both the EPA-funded process and COTE. In the end, Portland's representative decided not to sign on. While a different representative of the Bureau of Environmental Services participated in training sessions explaining how market tools can be used, the City has taken no further steps to purchase or develop ecosystem services credits under the Willamette Partnership's marketplace or any other trading regime.

Portland's unwillingness to enter the marketplace is somewhat surprising given the city's history of investment in the creation of ecosystem services. In fact, in the December 2008 edition of the American Planning Association's magazine, Portland was described as one of four American examples of successful investments in *green infrastructure*, defined as

> the interconnected network of open spaces and natural areas—greenways, wetlands, parks, forest preserves, and native plant vegetation—that naturally manages storm water, reduces the risk of floods, captures pollution, and improves water quality. (Wise 2008, 1)

In short, Portland invested heavily in the development of so-called *green infrastructure*, a form of infrastructure explicitly designed to produce ecosystem services, yet declined to enter the market, either as a buyer or seller of ecosystem service credits. Given Portland's importance in the Willamette River basin, economically, environmentally and politically, as well as its reputation as an environmental leader, it is important to understand the city's reservations, especially since the city's view reveals underlying concerns about the role of place and investments in place in an ecosystem services market.

A set of interlocking concerns drove the city's decision. These began with worries about losing environmental goods and having them replaced by restoration activities in rural areas far from the city:

> I will tell you [that] one of the big reasons why the City also declined to sign it [the final agreement of Counting on the Environment] was they were unsure. It was the offsite mitigation requirement. The city wanted to be focusing their mitigation and protection and restoration requirements within the city and not having them shipped out to these pilot sites that were outside of city boundaries. (Interview with author, August 5, 2011)

This statement reflects a concern about the scale at which environmental mitigation should take place. Some people in Portland argued that environmental restoration should occur near the location of whatever environmental destruction was happening. This concern is not unjustified. The creation of markets for wetlands repeatedly illustrates a pattern of replacing urban with rural wetlands (BenDor and Brozović 2007; BenDor, Brozović and Pallathucheril 2007; King and Herbert 1997; Ruhl and Salzman 2006; Womble and Doyle 2012). The logic is purely economic: developers prefer to build (and thus drain or fill wetlands) in urban areas with higher land values, and then create wetlands for mitigation purposes in more rural areas, where land values are lower and connectivity to larger green spaces can be established more easily.

So Portland's reluctance to compensate the destruction of ecologically marginal green spaces in urban areas with significant environmental improvements in rural areas, even at lower cost, is directly linked to the meaning of place. The assessment of the responsible city officials was that residents of urban areas, at least in Portland, value proximity over efficiency. They like green spaces in their city for reasons not captured by a guarantee that the ecosystem services they stand to lose will be provided somewhere else:

> We have a larger population that is beginning to become more vocal about wanting to see green and sustainability and fish and wildlife in these urban areas. (Interview with author, August 5, 2011)

Such concerns can be understood as the inverse of what planning scholar Sarah Dooling has called "ecological gentrification" (2009). This is a process by which environmental amenities, specifically urban green spaces, are increasingly concentrated in wealthy areas or where access is restricted for low-income residents, especially the homeless. In this case, a reluctance to participate in the ecosystem services market is stimulated by a fear of something that we might call "ecological outsourcing." It is unacceptable to city residents and the officials who represent them to lose the benefits of a valued and proximate ecosystem. For example, the water-cooling effect of a tree can be "enjoyed" at any point downstream from that tree, but to sit underneath the tree and enjoy the shade it casts requires being right next to it. When water temperature reduction is "outsourced" to an area outside the city, residents are effectively prevented from enjoying the ancillary benefits of the trees near them.

This unwillingness of NMFS and one of the "greenest" cities in the country to participate in the implementation of an ecosystem services market at the watershed scale highlights the challenge of taking account of concerns about place. The fundamental logic of market-based thinking offers no solution to this problem. Both NMFS and Portland's Bureau of Environmental Services participated actively in the preliminary negotiations, but in the end, their concerns about salmon and the needs of city residents could not be assuaged by the promise that bureaucracy would be cut back and economic efficiency enhanced.

2.c Moving Matter out of Place in Pennsylvania

Many environmental concerns in the Chesapeake Bay watershed revolve around nutrients such as phosphorous and nitrogen entering the Bay and causing algae blooms and *hypoxia*, a shortage of oxygen in the water that can kill great numbers of fish. Agricultural activities throughout the watershed, especially those that cause fertilizer and manure to enter the streams that eventually flow into the Chesapeake Bay, have significant cumulative impacts on water quality in the Bay. This dynamic is well known, and efforts to clean up the Bay have long focused on agricultural activities in rural areas far upstream of where the Potomac and Susquehanna enter the Atlantic Ocean (Dauer, Ranasinghe and Weisberg 2000).

The Commonwealth of Pennsylvania was first in the Chesapeake Bay watershed to develop and implement a Water Quality Trading scheme. Its goal was to improve water quality in its rivers, as well as in the Chesapeake Bay. Water treatment utilities and factories that discharge water into rivers and streams are allowed to achieve water quality standards in multiple ways. One involves a nutrient trading scheme between point sources and non-point sources. The former are facilities with distinct outlets, such as water treatment facilities and industrial water users, that discharge water into streams. Government agencies can set a quantitative standard and allow point sources to trade credits with each other. As one facility decides to upgrade its technology, it can swap the value of the pollution reduction it produces with a plant that is not yet ready to make such an investment. This falls squarely under the logic of *pollution markets*.

Non-point sources refer to diffuse discharges of water into streams, like run-off from a farm field after rainfall. That water can contain pollutants from a whole variety of sources, like fertilizer, pesticides or manure. Market schemes that allow trading between point sources, which need credits, and non-point sources, which can generate credits by planting riparian buffer zones or grass strips between crops, follow the logic and promise of ecosystem service markets.

Pennsylvania's market allows point sources to buy credits from each other, and from non-point sources as well. A central concept in the development of non-point source credits is the idea of "Best Management Practices," or BMPs. These are expected to reduce levels of nitrogen and phosphorous entering the water. In Pennsylvania, 15 BMPs were approved. Implementing one of these can generate credits for landowners. One of the most popular practices in Pennsylvania was known as the aforementioned *manure transport*. This consisted of transporting chicken manure out of watersheds connected to the Chesapeake Bay, namely the Potomac and Susquehanna River basins, into the Ohio River basin, which eventually drains into the Gulf of Mexico via the Mississippi River. This practice quickly became very popular among landowners, but raised deep concerns among agency staff members and environmental advocates. The practice was effectively removed from the list of approved Best Management Practices after a few years. The development of, resistance against and eventual ban on manure hauling provides a final example of the discomfort with displacement that has emerged in response to ESMs.

The first market in the Chesapeake Bay watershed

The World Resources Institute (WRI) published an influential report proposing Water Quality Trading in the Chesapeake Bay watershed as early as 2000 (Faeth 2000), although this was not the first time such an approach was suggested (Stephenson, Kerns and Shabman 1995). Elsewhere in the United States, local experiments with Water Quality Trading began as early as the 1980s. Many focused on trades between point sources (McGinnis 2001). The Environmental Protection Agency developed its draft framework to support watershed-based trading in 1996, and encouraged pilot projects anywhere in the nation. An early review of those efforts, titled "Will Nutrient Trading Ever Work?" counted 37 proposed trading schemes and revealed that only three included point source and non-point source trading (King and Kuch 2003). In the wake of this mixed review of pilot programs, the EPA released its formal Final Water Quality Trading Policy in 2003.[16] WRI began working with the Commonwealth of Pennsylvania to develop water quality markets for nutrients shortly after this policy was issued.

It took until 2006 for Pennsylvania to release its Trading Guidance[17] specific to the Commonwealth. This followed an extensive stakeholder participation process[18] that included more than 30 representatives of environmental organizations, municipalities and state agencies, broken down into various working groups. These committees focused on issues ranging from legacy sediments[19] to baseline determination. They typically met two to three times during the summer of 2006 leading up to publication of the guidance document. WRI's efforts focused on developing Nutrient Net, a web portal that enables

16 For the complete policy, see: http://water.epa.gov/type/watersheds/trading/finalpolicy2003.cfm Last accessed on 05/18/2015.

17 The basic goal of this guidance, and most others like it, was to inform the citizenry about the specific way an agency, in this case PADEP, interprets and plans to implement formal policy in practice, meaning Pennsylvania's Water Quality Trading Policy here.

18 For extensive summaries of the workgroup meetings, the listening sessions and the guidance itself, see: http://www.portal.state.pa.us/portal/server.pt/directory/chesapeake/8051?DirMode=1 Last accessed on 05/15/2015.

19 This term refers to sediment that has accumulated over decades, sometimes centuries, behind dams and in constructed ponds. After the removal of many of those dams, the sediment largely remains in place in stream banks, trapping phosphorous and nitrogen and sometimes other pollutants like heavy metals. River restoration activities that increase flow or generate more fluctuation in water levels often result in the erosion of those stream banks, releasing large quantities of those pollutants. For a description about legacy sediments in Pennsylvania, see: http://www.bayjournal.com/article/legacy_sediments_may_pose_threat_to_bay_cleanup Last accessed on 05/19/2015.

Table 2.2   Statutes and regulatory agencies in the Chesapeake Bay Bank

Credit Type	Regulatory Driver	Agency
Carbon Dioxide	Regional Greenhouse Gas Initiative	Energy and environmental agencies in participating states
Bog Turtle Habitat **Trout Habitat**	Endangered Species Act, Sections 7 & 10	U.S. Fish and Wildlife Service
Forest	Maryland Forest Conservation Act	Local Jurisdictions (counties/ municipalities)
Water Quality— Nitrogen **Water Quality— Phosphorous**	Clean Water Act, Section 303(d)	U.S. Environmental Protection Agency
Wetlands	Clean Water Act, Section 404	U.S. Army Corps of Engineers

Source: Kane and Prendergrass (2007); Sprague, Price and Remuzzi (2007)[20]

landowners to calculate, register and sell credits, and permits entities to post bids and buy credits to offset their impacts.

Even more ambitious than WRI's work on Water Quality Trading in multiple states in the Chesapeake Bay watershed, the Pinchot Institute began to develop an approach to integrate multiple credit types into a single ESM in the Chesapeake Bay in 2007. Water Quality Trading was initially proposed as an important component of this new ESM, but far from the only credit type to be traded. The Chesapeake Bay Bank, as this new ESM was to be called, would incorporate all three market mechanisms under federal law and two additional credit types, for carbon dioxide and forestlands (Table 2.2). The specific regulations that would require power plants to buy credits for carbon dioxide and developers to buy credits for forestland were the Regional Greenhouse Gas Initiative (RGGI) and the Maryland Forest Conservation Act, respectively. In theory, restoring forests by planting trees could generate both of these credit types. Maryland, Delaware and New York are the only states in the Chesapeake Bay watershed actively participating in the RGGI. Maryland's Forest Conservation Act is obviously only applicable in that state. This would mean that potential sellers of these credits would have to develop restoration or conservation projects in the relevant jurisdictions.

As was the case in Oregon, it was difficult to move from designing the basic outlines of a market on paper to actually generating credits and

20 Presentation titled "Ecosystem Markets in the Chesapeake: The Bay Bank." Available at: http:// www.katoombagroup.org/documents/files/Sprague.pdf.

managing transactions. To appreciate the scale and level of ambition of the Chesapeake Bay Bank proposal, it is instructive to take a closer look at the challenges of developing Water Quality Trading, only a single credit type in one state, namely Pennsylvania. The potential credit buyers in the Pennsylvania market are the more than 800 large (municipal) sewage treatment facilities and industrial wastewater treatment plants. Since 2010, they are required to reduce their emissions in line with limits specified in each of their NPDES permits. By reducing emissions beyond the limits stipulated in their permits, they can also become credit sellers. The Bay-wide TMDL had not yet been issued in 2006. This created some doubt about the overall viability of the market since it was based on water quality standards that exclusively applied to and were effectively enforced by the Commonwealth of Pennsylvania.

To generate credits, landowners could implement one of more than 15 Best Management Practices. These reduce or eliminate nutrient and/or sediment runoff and consist of practices ranging from the creation of riparian grass buffers to the permanent installation of mortality composters.

Each BMP is associated with an effectiveness estimate or efficiency level specifying the expected reduction in nutrient or sediment runoff that the BMP is intended to generate. For example, the mortality composter[21] is associated with a nitrogen efficiency of 14%. This means that composting carcasses in a controlled facility, using manure to aid in the process and applying it to the land as a source of nutrients is presumed to generate a 14% reduction in nitrogen runoff. These BMPs and their associated efficiencies are used to calculate the reduction in nutrient runoff on a specific farm or field to produce an exact number of credits. These credits are then expressed in pounds of nitrogen or phosphorous, per year, that will not reach the Chesapeake Bay as a result of BMP implementation.

The complexity of credit calculation was an important barrier to getting the market started. The logic of an ecosystem service market requires credits that are measurable and affordable, or at least more affordable than infrastructure upgrades like new filters or storage basins. Just as decision makers tried to reduce transaction costs by increasing the scale and scope of the markets for ecosystem services in the Willamette River basin, Pennsylvania's Department of Environmental Protection was looking for ways to make the market work by generating trades and reducing transaction costs. An agricultural consulting company, Red Barn, suggested a new way to reduce runoff.

---

21 Pennsylvania's list of approved and pending BMPs and their efficiencies can be found here: www.portal.state.pa.us/portal/server.pt/document/bmpdescriptions_pdf Last accessed on 05/22/2015.

Red Barn's CEO described his discussion about trading with Department officials:

> We came to this point of, well all right, what if we do this or if we do that BMP, like what about fencing or what about riparian buffers? All of this stuff was really hard, and no science to back it up. We were sitting down and were like, piles of poultry manure, what if we took that out? (Interview with author, January 30, 2013)

Red Barn proposed to truck poultry manure out of the impaired watershed so it could be deposited elsewhere, a BMP that became known simply, if unpoetically, as *manure transport*. Pennsylvania's DEP formally describes this BMP as: "Transport of livestock manure from areas of high concentration to areas of low concentration, or the transport of manure out of the Chesapeake Bay watershed."[22]

The relative ease with which the reductions could be measured, simply by weighing the trucks as they left the watershed and conducting a basic chemical analysis of the manure to determine nitrogen and phosphorous levels, was a key reason this BMP represented a breakthrough. Red Barnes' CEO vividly remembers the reaction from state officials at the time: "That sounds great; you just can't take it east, you have to take it west" (Interview with author, January 30, 2013). This proposal about where to take the manure, as well as the lack of oversight over exactly where the manure was transported, proved important.

At the time, given the enthusiasm for trading, it is not surprising that Pennsylvania observed a quick rise in the number of credits generated under the new manure transport BMP. Between 2006 and 2010, the overwhelming majority of pollutant reduction projects in Pennsylvania's market consisted of poultry manure export. Of the 52 projects that the Pennsylvania Department of Environmental Protection certified before 2010, 44 were based on simply trucking chicken manure from the impaired river basins to the Ohio River basin, with little publicly available documentation about where that manure was deposited.[23] This practice was immediately controversial. The EPA and an environmental advocacy group, PENN Future, ultimately challenged it.

---

22 Pennsylvania Department of Environmental Protection, list of approved BMP descriptions for the Chesapeake Bay Watershed model, available at: http://www.dep.state.pa.us/river/nutrienttrading/calculations/docs/BMPDescriptions.pdf Retrieved on 03/04/2014.

23 The overview of credited projects available on PADEP's website does not indicate where exactly the manure is deposited, and the "Nutrient Trading Evaluation Report" commissioned by PENN Future indicates that little to none of this information is publicly available. The full report is available at: http://www.pennfuture.org/UserFiles/File/Water/RespFarm/Report_NutrientTrading-Eval_20110919.pdf. Retrieved on 07/13/2014.

Poultry manure and the boundaries of trading

Hauling manure out of areas with excess had been a subsidized practice in Delaware and Maryland long before the creation of any trading scheme in the Chesapeake Bay. The EPA was on record disapproving this method of addressing nutrient problems in the Chesapeake Bay: "Ultimately, applying manure on agricultural land is a quick fix, but not a long-term solution. You have stricter regulations coming down, you have reduction in land, and eventually your soils are going to become phosphorus saturated" (Gary Shenk, Chesapeake Bay Program, quoted in Blankenship 2005). But the creation of the trading scheme in Pennsylvania contributed to the popularity of manure transport, at least in that state.

In June 2010, PENN Future, an influential advocacy organization, released a technical assessment of the nutrient trading program. The report was clearly negative about the viability of manure transport as a strategy for addressing water quality problems. Over time the export of nutrients to other basins was likely to adversely affect water quality there as well. Referring to the transport of manure to the Ohio River, which flows into the Mississippi, the report stated that there were already serious hypoxia concerns in the Gulf of Mexico. PENN Future also argued that the EPA had made clear that water quality improvement efforts were likely to begin sooner rather than later in those parts of Pennsylvania where creeks and streams that eventually drained into the Gulf of Mexico were located (Century Engineering 2011, 26–27).

In October 2010, a few months after the release of this highly critical report, Pennsylvania's official nutrient credit trading regulation came into effect.[24] Coinciding with this new framework, a new calculation method for credit generation was introduced. It dramatically reduced the profitability of manure transport. While much of the transported manure counted directly toward credit generation, this new method required attention to the attenuating effect on the water the nutrient would seep into. This meant that nutrients removed from land further away from the Chesapeake Bay produced fewer credits. After a few years of rapid growth, the introduction of this new credit calculation method in Pennsylvania dramatically reduced the number of "manure transport" credits generated.[25] Regardless of this decline, manure

---

24 The regulation is 25 Pa. Code § 96.8, titled "Use of offsets and tradable credits from pollution reduction activities in the Chesapeake Bay Watershed," and is available at: http://www.pabulletin. com/secure/data/vol40/40–41/1927.html Last accessed on 05/21/2015.

25 In 2011, six manure transport projects were credited by PADEP, out of a total of 26 pollutant reduction activities. In 2012, 56 projects were credited, none of which included manure transport. See: http://www.dep.state.pa.us/river/nutrienttrading/projects/index.htm Last accessed on 07/ 14/2013.

transport projects continued, and credit developers like Red Barn hoped that a potential drop in fuel prices would make the BMP more profitable again.

Later in 2010, the EPA released its pollution standard, or TMDL, for the entire Chesapeake Bay.[26] This meant that Pennsylvania's trading program now fell directly under the EPA's authority and would be reviewed under a set of guidelines described in an appendix to the TMDL.

This led to an EPA review of all trading programs in the Chesapeake Bay watershed. The outcome was that Pennsylvania's Department of Environmental Protection no longer supported manure transport as a creditable BMP. Pennsylvania DEP's exact reasoning is not fully transparent. In the EPA's formal review of Pennsylvania's trading program, manure transport is not mentioned.[27] PADEP organized a meeting with stakeholders to explain the outcomes of the EPA's review and the changes that would be made in the trading program, along with enhancements that PADEP was likely to implement as a result. At this meeting, or *enhancement workshop*, the following decision was made, as described by Red Barn Trading's CEO:

> At this enhancement workshop, which was held by [Pennsylvania] DEP, it was in the PowerPoint, a thing that said no more manure hauling. (Interview with author, January 30, 2013)

There was little further explanation of why this decision had been reached, but he went on to describe his own reaction to this announcement:

> I understand it, and that will just drive the market, or us to come up with newer and better and more [defensible] [BMPs], even though I think I can defend it all day and be quite serious about it. But that does not mean it [the manure hauling] is the right thing to do. If there is enough momentum that says that it's wrong, I'm quite willing [to stop]. (Interview with author, January 30, 2013)

His mention of momentum refers to the repeated criticism of this BMP by EPA officials, along with those from other states, at meetings and workshops about nutrient trading in the Chesapeake Bay. He was not surprised by the "notoriety" of manure hauling in the community working on nutrient trading in the Chesapeake Bay: "Do I understand all of the criticism of it? Of course!" This understanding reveals that the developer of the manure transport BMP

---

26 See: http://www.epa.gov/chesapeakebaytmdl/ Last accessed on 05/19/2015.

27 The EPA conducted a review of all the trading programs in the Chesapeake Bay states in 2011. The results of this review, as well as the responses by the states, can be found at: http://www.epa.gov/reg3wapd/pdf/pdf_chesbay/Phase2WIPEvals/Trading_Offsets/PortfolioOfReports.pdf. Last accessed on 05/20/2015.

was aware that, while hauling manure was perfectly in line with the market logic prevailing within PADEP, it was not acceptable to those outside that circle. A second point that emerges from his description is that he fully expected the same dynamic to persist, with credit brokers coming up with *newer and better and more defensible* BMPs and others either supporting or attacking them.

This brief overview of the contentious history of a single BMP shows that the challenges associated with place in the creation of ecosystem service markets were not limited to the moral imperative to protect unique places, or the desire to facilitate proximity to green spaces for urban dwellers. While *moving matter out of place* was politically acceptable in Pennsylvania for a few years, it was never accepted as a BMP in Maryland or Virginia, and was ultimately denied in Pennsylvania as well. This suggests that reactions to displacing *sociospatial arrangements* can lead to varying market-making practices between states and over time.

The challenge of incorporating concerns about place into the Water Quality Trading program in Pennsylvania did not result in a collapse of the market. Pennsylvania's nutrient trading program remains the most active water quality market in the Chesapeake Bay watershed, based on the volume of trades.[28] Between 2010, when manure transport was halted, and 2015, regular auctions of nutrient credits took place with prices that ranged from $1.22 to $4.00 per pound of avoided nitrogen or phosphorous.[29] In the Chesapeake Bay watershed, Pennsylvania's market is considered relatively well established and remains significant in size.[30] But the manure transport BMP is frequently mentioned as an obstacle to the potential integration of the emerging nutrient markets in Maryland and Virginia with Pennsylvania's. As such, this BMP has become an example of the risks associated with following the desire to reduce transaction costs to its logical extreme.

Manure hauling and its sudden discontinuation made it more difficult for advocates of Chesapeake Bay-wide trading of environmental credits to implement their visions. While WRI has continued its efforts to develop Water Quality Trading schemes in various states in the Chesapeake Bay watershed, the proponents of the Chesapeake Bay Bank largely dropped water quality credits from their list. The latter's efforts have focused more narrowly on the development of software to make forestland conservation in Maryland and

---

28 For the details on the publicly available trades, see: http://www.markit.com/en/products/environmental/auctions/pennvest.page. Last accessed on 05/20/2015.

29 This was calculated using the registry: http://www.markit.com/en/products/environmental/auctions/pennvest.page. Last accessed on 05/21/2015.

30 The credit-generation projects that have been certified by the Department of Environmental Protection can be found on its website at: http://www.portal.state.pa.us/portal/server.pt/community/nutrient_credit_trading/19518. Last accessed on 05/22/2015.

species habitat banking more attractive to landowners. The contestation over manure hauling in Pennsylvania highlights the challenge place presents to ESMs and raises questions over the practical feasibility of integrating multiple credit types into a single marketplace at the scale of a watershed the size of the Chesapeake Bay.

2.d Markets as Displacements

Moving environmental goods, such as urban green spaces in Portland, and environmental bads, such as poultry manure in Pennsylvania, may make sense in ecological terms. Such moves may make even more sense in economic terms, but these kinds of actions undermine carefully managed socio-spatial arrangements or *places*. While the goals of ecosystem service markets in general and in the Chesapeake Bay watershed and the Willamette River basin in particular are rather broad, the underlying logic of markets makes it difficult to take place-specific concerns into account. These examples show why attempts to create places for ecosystem services markets, both in the Willamette River basin and in the Chesapeake Bay watershed, have met significant resistance.

The implementation of markets results in a transfer of environmental goods or bads from one place to the other. In Pennsylvania's manure hauling example, this happened in a very physical sense, but Portland and NMFS's refusal to even enter the marketplace shows that the possibility alone can raise barriers to market creation. The experiences in Oregon and Pennsylvania show that the movement toward ESMs or the installation of Best Management Practices should be evaluated from more than a purely ecological point of view. Political and social effects are associated with market creation that make for significant challenges to the fundamental premise of these markets: that displacing certain (parts of) ecosystems can result in broad gains, which reduce the tension between environmental and economic goals.

If the proponents of markets for ecosystem services want to create durable institutions that can generate the support they need from the communities involved, they will need to take into account the connections people feel to the places in which they live and work. Perhaps what is surprising is that questions about place, for example, which places should be excluded from any market activity or the spatial distribution of ecosystem services, were not addressed more carefully during the design of these new institutional arrangements.

Place has received insufficient consideration in ESMs, in part because these endeavors have been pushed by the logic of the market that views trade-offs as always possible and ratios and priority areas as sufficient to buffer any lingering concerns. Portland and NMFS did not agree with this optimistic assessment. Pennsylvania DEP's reaction to complaints about the role of places or displacement was different. The agency adjusted its calculation method to reduce the appeal of manure transport. This response highlights the second important element, and the source of the second key challenge, of markets for ecosystem services, namely the measurement systems that determine the exact quantities and qualities of the credits at the center of these markets. It is this second challenge I turn to in the next chapter.

Chapter 3

PRODUCING EQUIVALENCE

In the summer of 2011, the Willamette Partnership organized a series of training sessions on how to calculate the number of credits a particular environmental restoration project should receive. During this workshop, two wetland ecologists engaged in a protracted debate about whether beavers were present on a site, and if so, how likely they were to prevent certain species of trees from growing. While this was only a training session, it was obvious that different answers to these questions would profoundly impact the number of credits a restoration project on this site could generate. Tree growth is the key determinant in whether water-cooling shadows are produced, but the Shade-a-Lator does not include the presence of beavers in its calculation. Even a single pair of beavers can quickly reduce the number of standing trees, and therefore the amount of shade, in a riparian zone.

This conversation highlights common uncertainties associated with environmental restoration work, but it also reveals a more specific problem the developers of spreadsheets like the Shade-a-Lator have to overcome. For an ESM to function, its proponents have to produce compelling evidence that newly created ecosystem services in one place are equivalent to or greater than a negative environmental impact elsewhere. Each effort to create an ESM therefore needs to develop or select a way to calculate the number of credits a landowner receives for a restoration project. Relevant tools include spreadsheet-based calculators, but also elaborate rules about site visits to visually confirm model outcomes and training programs to ensure that capable experts conduct those site visits. I use the phrase "measurement systems" to describe this complex of models, protocols and practices, which together produce the quantity of credits a participant in an ESM needs to buy or is able to sell. These measurement systems are at the heart of the second challenge for ESMs, namely that of producing equivalence.

It is easy to disagree about a calculation of how many trees need to be planted to compensate for a specific quantity of discharged warm water in a different location. Two experts can argue, at length, about whether beavers are present on a site, and how their presence would affect the survival rate of newly planted trees. But in a market, it is crucially important to establish broad trust in the "currency," like *kilocalories per day* or *pounds of nitrogen per year*. This kind of has to be accurate, but it must also be in line with relevant legislation and jurisprudence. In case of litigation, all too common in the environmental realm, the calculation method and its outcomes have to be robust enough to be considered legitimate in the courts (Jasanoff 1995). In an ESM, measurements inform tangible and financially significant decisions, for example, about whether to build a costly chiller or invest in planting trees for a 20-year period. A large circle of people, like regulators, landowners, utility managers and environmental advocates, has to be familiar with and willing to abide by the outcome of such calculations.

The question of what type of measurement system is good, or even acceptable, turns out to be difficult to answer in theory and in practice. A review of knowledge systems at the intersection of society, environment and economics argues that these measurement systems have to produce information that is salient, credible and legitimate (Cash et al. 2003). These criteria certainly apply to ESMs as well, but are not necessarily sufficient. The Willamette Partnership, an influential market proponent in the Pacific Northwest, has produced multiple reports specifically on its experiences developing these measurement systems. Its conclusions include the need for these calculation methods to be "usable" (Cochran and Robinson Maness 2011, 23) and "trusted" (Sanneman, Culliney and Cochran 2014, 4). All these criteria relate to the sociopolitical, as opposed to the narrow technical, context within which these measurement systems are evaluated. But figuring out how to develop measurement systems that meet these criteria is not easy.

Two closely related types of tension frequently emerge in the development and evaluation of measurement systems. The first centers on the different interests of key participants in ESMs. The second is that these measurement systems relate to very different knowledge systems, or epistemologies. Before turning to the case studies of measurement systems in ESMs, I'll describe these common sources of tension in more detail.

The role of equivalence

In an ecosystem services market, the primary actors in any transaction frequently have quite different interests in how the measurement system operates. Yet the overall functioning of the marketplace relies on the belief that the value of the ecosystem services is the same or higher when compared to the cost of the activity harming the environment, whether that activity is discharging warm water or destroying a wetland. Chapter 2 shows how one element that makes it difficult to convince people of the possibility of equal value is the discomfort with market-driven displacement. But even willing potential participants in ESMs frequently disagree about the correct approach to assessing the quality and quantity of credits an ecosystem restoration project generates. The basic structure of a market creates some of this tension, and the very disparate activities connected through the market, like wastewater treatment and tree planting, exacerbate it.

The "buyers" of credits, often large institutions like wastewater treatment utilities, energy companies or real estate development companies, require a way of calculating the credit production of a restoration project that provides a high level of certainty that by buying credits they will meet their legal obligations. For example, a wastewater treatment utility that invests millions of dollars in riparian restoration needs its investment to compare favorably to installing new chillers or nutrient filters in terms of the risk the organization is taking on. Since many credit buyers are required by law to purchase credits to retain or get a permit, buyers need to know that by buying particular credits they will meet their legal obligations. When a cooling installation at a water treatment facility does not operate according to specification, or completely breaks down, it is likely that the utility will quickly notice this and get it fixed. But when a storm downs a number of trees next to a small creek, the utility is far less likely to notice this or be able to quickly intervene. Yet, at least in principle, both situations could jeopardize the utility's compliance status with federal and state regulations, and open up the possibility of fines or litigation. So the measurement systems' calculations, and the assumptions they build on, need to provide a high level of regulatory certainty for the buyers.

The civil servants tasked by the federal or state government with enforcing these requirements, typically at the state and federal levels, also have to be sure that credit calculations reflect meaningful progress, in terms of improved water quality, species' survival or the like. This means that these regulators are likely to push for ways of measuring such progress, including regular site visits. Frequently, market-based approaches to regulatory compliance are the subject of close scrutiny by environmental groups[3] who worry that reliance on markets is merely a way to relax environmental protection standards. So they too want to know that appropriate objectives are being met. State and federal agencies often view rigorous and detailed monitoring systems, developed by well-known experts in relevant scientific fields, as the best way of demonstrating that they are doing their jobs.

The sellers of credits, typically landowners in more rural areas, want to be certain that they can sell the issued credits based on the restoration efforts they

---

3 See for examples relevant to the efforts in Oregon and the Chesapeake Bay states: http://northwest-environmentaladvocates.org/water-quality-trading-innovation-or-hoax/ and http://www.foodand waterwatch.org/pressreleases/nearing-40th-anniversary-of-the-clean-water-act-advocacy-groups-challenge-national-water-pollution-trading-model/. Both last accessed on 07/01/2015.

make. This means they will want performance review metrics that are easy to use, since the onus is typically on the credit developer to prove that restoration projects qualify. Many landowners are not enthusiastic about frequent visits from regulators or credit verifiers. This means that a virtual credit calculation system, that uses information provided by farmers and existing data sources like aerial photography are likely to be most popular among this group.

These three broad categories of actors—buyers, regulators and sellers—all have different concerns about the performance of the measurement systems used to calculate ecosystem services production. These actors also evaluate the performance of those measurement systems in different terms and based on different criteria. In practice, this makes it complicated to develop a system that meets all of their concerns. What exacerbates this challenge, and significantly raises the stakes for all three groups, is the fact that once investments are made and regulatory approval is given, continued support for the way measurements are made is necessary to maintain the viability of the trading system. Many of the contracts between landowners and credit brokers or regulated entities like water treatment plants are for long periods of up to 20 years. Once a credit has been approved, the underlying measurement system has to remain operational for the duration of that restoration project.

Science, markets and the law

A second frequent challenge in reaching agreement about what makes for a good measurement system is not necessarily based on interests, but more broadly on what counts as meaningful knowledge. In order to calculate ecosystem services so that credits can be traded, it is rarely enough to choose among existing methods and tools developed by academic scientists. While the growth of ecological economics as an academic discipline[4] has produced an array of

---

4 As I described in the first chapter, the history of ecological economics as an independent field is roughly three decades old, and closely tied to key people like Herman Daly, Richard Norgaard, Gretchen Daily and Robert Costanza. Daly's work is well known for its emphasis on the flaws of Gross Domestic Product as an economic indicator, so many early measurement systems emerging from this line of work were intended to replace those, like an Index of Sustainable Economic Welfare (ISEW). His 1997 book *Beyond Growth: The Economics of Sustainable Development* remains perhaps his most widely read publication, but this index was first published in his 1989 book *The Common Good* and continues to influence efforts to implement "green accounting" for governments. The following decade saw a shift to valuation techniques, many of them based on replacement cost or willingness to pay surveys, often for specific ecosystems. The most prominent study of this kind is Robert Costanza et al.'s oft-cited "The Value of the World's Ecosystem Services and Natural Capital," published in the journal *Nature* in 1997. Since the formal, practical uptake of these approaches by national governments remained limited, as Gretchen Daily described in a special issue of the Proceedings of the National Academies of Science in 2008, the most recent decade has seen an increase in attention to mechanisms to encourage direct payments for ecosystem services (PES). Here, the

measurement technologies, the systems used in ESMs have typically required modification to produce outcomes appropriate to each market. This places these tools squarely in the realm of *regulatory science* (Jasanoff 1990), meaning that the constraints under which they were developed did not match idealized norms of scientific independence and fundamental inquiry. Instead, experts create these models and techniques to solve specific problems at a particular point in time, and in this process social and political judgments factor into design decisions. Environmental complexity and the constraints of the regulatory environment are not the only challenges for the designers of ESM measurement systems.

Geographer Morgan Robertson has described the function of such measurement systems in terms of producing "nature that capital can see" (Robertson 2006). In his view, ecosystem services metrics and the ecological functions they make visible exist at the nexus between science, state and market. Not only do these measurement systems need to meet scientific and regulatory criteria, they also need to function effectively within a market context. This provides an additional set of challenges and demands.

To highlight the precarious place of metrics caught between science, policy and economics, Robertson describes numerous interactions between members of wetland review teams and "bankers," in which they, often jokingly, disagree about the presence of a specific plant or critter, since it might make the wetland more valuable in ecological as well as economic terms. The presence of a specific plant in the wake of a restoration effort could increase the number of credits allowed and therefore the amount an owner can making selling the credits. At the same time, the relative value, expressed in the number of points allotted for a specific species, is often influenced by its presence on a list of rare, threatened or endangered species. These lists reflect state or federal policies and not just scientific judgments.

The divergent logics of science, state and market, when they come together in a measurement system, can result in a lack of stability. Not only does a measurement system need to meet the requirements flowing from these different domains, their outcomes are likely to be challenged based on one or more of the logics that apply in each domain. Regulators, advocates and credit buyers and sellers have challenged the measurement systems in Oregon, the Ohio River basin and the Chesapeake Bay on all of these grounds, often simultaneously, meaning that the same system was considered not scientific by one group, yet not in line with relevant jurisprudence by another.

measurement systems are focused on quantifying a specific service, and its price is not calculated, but determined through negotiation. In that same year, a special issue of the journal *Ecological Economics*, edited by Sven Wunder, Stefanie Engel and Stefano Pagiola, was dedicated to PES systems and the measurement technologies associated with them.

The history of the measurement systems in markets for ecosystem services reflects the complicated regulatory, political and economic environment in which measurement systems have to function, as well as the fact that different efforts under the broad rubric of creating ESMs have led to different approaches to measuring ecosystems services. The increasing diversity of these hybrid institutions has not gone unnoticed in the academic and professional communities that follow the development of ESMs. Repeated calls for standardization have been voiced (Boyd and Banzhaf 2007; Willamette Partnership, Pinchot Institute for Conservation and World Resources Institute 2012). The reasons such standardization is difficult to achieve, however, become apparent if we look in more detail at the adjustments that have been made in trying to make these systems work in practice in the Midwest and Oregon.

3.b Pipes versus Fields in the Ohio River Basin

To fully appreciate the difficulties associated with producing equivalence between polluting activities and the creation of ecosystem services, it is instructive to analyze the Electrical Power Research Institute's (EPRI) effort to create an ESM in the Ohio River basin. EPRI is a research organization created to support the electricity industry, primarily but not exclusively in the United States. Its membership largely consists of electricity companies and utilities, which together represent a large portion of the American energy market.[5] EPRI has been interested in the potential of ESMs to mitigate the overall environmental impact of electricity generation for a long time. First, many utilities in the United States have been familiar with the basic idea of environmental markets since the Clean Air Act Amendments of 1990, which resulted in the creation of a market in air quality credits. Second, most large-scale electricity generation not only emits pollutants into the air but also requires significant quantities of water, for example, to create the steam that drives turbines and for cooling and filtration purposes. This water is discharged into streams and rivers after use, typically with elevated levels of a range of components, including nitrogen (EPRI 2006).

EPRI's experience with trying to create an ESM for nutrients and greenhouse gasses in the Ohio River basin reveals two important barriers related to measurement for ESMs. The first obstacle is that the differences between wastewater discharged by a power plant and water quality improvements achieved on farmland cannot necessarily be resolved by creating ever more detailed measurement systems. The second barrier is that the emergence

---

5 See: http://www.epri.com/About-Us/Pages/Our-Story.aspx. Last accessed on 08/30/2015.

of multiple approaches to measuring highly similar ecosystem services can threaten to destabilize the basic credibility of the ESM approach.

A business case

One of the first formal statements of interest from EPRI regarding Water Quality Trading occurred in 2002, when EPRI published a detailed report of market design issues and options (EPRI 2002). This preceded the EPA's release of its formal guidance on Water Quality Trading and came well before early examples of Water Quality Trading like the shade trading program in the Tualatin River basin in Oregon received some acclaim. In 2005, EPRI decided to not simply follow and report on the development of ESMs, but established a Water Quality Trading focus group. This resulted in the decision to try to actively develop an ESM in 2006.[6] This effort became the Ohio River Basin Trading Project[7] and is now known as one of the most significant investments in ESM creation in the United States, with a strong focus on measurement and modeling of water and nutrients.[8]

From an environmental point of view, the Ohio River basin is a logical choice to try to create an ESM. The Ohio River is the largest tributary to the Mississippi River by volume and its watershed covers (parts of) 15 states, including Pennsylvania, West Virginia, Ohio, Indiana, Illinois and Kentucky. The Ohio River has an important impact on the quality of the water that eventually drains into the Gulf of Mexico via the Mississippi. Algae growth, fed by nutrients like phosphorous and nitrogen, has significantly reduced oxygen levels in the northern Gulf of Mexico, resulting in the so-called dead zone (Rabalais, Turner and Wiseman 2002). The Ohio River contributes a significant part of all the nutrients that flow into the Gulf just south of New Orleans, an estimated 36% of the nitrogen and 35% of the phosphorous (Alexander et al. 2008).

The initial choice for the Ohio River basin as the site for EPRI's attempt to create an ESM was deliberate and not simply based on these environmental grounds. In 2007, EPRI identified a very different selection criterion to select the location of its "demonstration project" (EPRI 2007a). EPRI's own research indicated that trading programs were successful in those places

---

6 For a detailed timeline, see the various newsletters EPRI published about this project since 2009, available at: http://wqt.epri.com/schedule-updates.html. Last accessed on 08/15/2015.

7 For up-to-date information on this ESM, see http://wqt.epri.com/index.html. Last accessed on 08/21/2015.

8 One example of recognition for EPRI's remarkable efforts is the fact that the project received the 2015 U.S. National Water Prize, handed out by the U.S. Water Alliance. See: http://uswateralliance.org/u-s-water-prize/winners/2015-prize-winners/. Last accessed on 08/21/2015.

where water dischargers (like power plants and water treatment facilities) were actively compelled to improve water quality by state and federal agencies under the Clean Water Act. In other words, where trading was directly linked to the National Pollution Discharge Elimination System (NPDES) permits based on TMDL allocations, permit-holders and landowners were more likely to buy and sell credits (EPRI 2007a, 4–4). That same report contains a long list of dormant or inactive markets, where people had tried to create the infrastructure that enabled trading, but no willing buyers emerged since there was no, or no impending, regulatory requirement to reduce the nutrient levels in the water.

Following this finding, EPRI selected three areas where it could experiment with an ESM by creating a demonstration project; the Chesapeake Bay watershed, the Catawba River basin (in North and South Carolina) and the Ohio River basin (EPRI 2007b). A central goal of this pilot was for an electricity-generating company with a substantial number of coal-fired power plants (since the filtration of that air produces high levels of nitrogen in the water that is discharged by the plant) to achieve pollution reductions through the purchase of credits developed by non-point sources, meaning farms or other large-scale land uses. A secondary goal was to simultaneously develop greenhouse gas emissions reductions, or "carbon credits" on agricultural lands as well, thereby creating not simply a Water Quality Trading scheme, but a more integrated market for multiple ecosystem services.

Besides the selection criterion of current or foreseeable regulatory limits on water pollution (through the Total Maximum Daily Loads, or TMDL process), EPRI reviewed the regulatory environment, level of interstate collaboration and a range of other aspects. Eventually, the Ohio River basin was chosen, even though specific pollution caps (in the form of TMDLs) had not been set in almost any of the large states in the watershed. Now, almost 10 years after EPRI decided to try to develop an ESM in the Ohio River basin, the importance of the absence of those specific pollution caps is difficult to overestimate. At the time of writing, none of the states has issued precise pollution caps, and most seem years away from doing so. This means that one of the key factors for success, identified by EPRI itself in 2007, continues to be absent in the Ohio River basin after almost a decade of work and millions of dollars invested.

At the time, the Ohio River basin seemed like a smart choice, given the existence of two well-known, but quite limited water quality credit–trading programs in Ohio. In one, a single cheese factory funded improvements on nearby dairy farms to reduce nutrient run-off into Sugar Creek (Moore, Parker and Weaver 2008). The other was larger in scale, both geographically and in terms of the number of credit buyers and sellers, in the Great Miami

River basin (Newburn and Woodward 2012). As a result of these programs, the relevant state and federal regulators were already familiar with the trading concept, if at a much smaller scale than what EPRI was proposing. Perhaps more important, these trading systems focused on a different type of waste-water discharger than power plants, namely municipal wastewater treatment plants and industrial water users.

Once the choice for the Ohio River basin had been made, EPRI worked to build connections to potential collaborators. First and foremost, this included its members in the region like American Electric Power (AEP), which owns 10 power plants in the watershed, and Duke Energy, owner of another three facilities in the Ohio River basin. In order to convince these electricity compa-nies to become credit buyers in a trading project that they were under no regu-latory obligation to participate in, EPRI published a feasibility analysis of the potential costs and benefits of supporting the development of an ESM in the Ohio River basin (EPRI 2010). This report mentioned the fact that most of the power plants in the region did not face limits on their nitrogen discharge at the time, but (in hindsight, incorrectly) anticipated that the responsible state agencies would set legal limits within the next two to three years.[9]

The business case for power companies to participate was therefore based on the idea that by supporting the development of trading as a compliance strategy, these firms could eventually meet the impending legal standards much more cheaply than by constructing capital upgrades to their on-site fil-tration capacity. In addition, early participation might give the power industry additional influence on critical regulatory decisions (EPRI 2010, 5–2).

Other important early partners took less convincing to get on board with the idea of an ESM in the Midwest. The Ohio River Sanitation Commission (ORSANCO), the U.S. Department of Agriculture (USDA) and the American Farmland Trust all supported EPRI's efforts. The fact that the energy industry, through its research organization EPRI, was interested in investing time and money in the creation of a large-scale ESM did not go unnoticed. In 2008, EPRI received a Targeted Watershed Grant from the EPA. Shortly thereafter EPRI also received a Conservation Innovation Grant from the USDA, mak-ing the total federal contribution to the project $1.3 million.[10] With funding

---

9 The report, titled "Program on Technology Innovation: Ohio River Water Quality Trading Pilot Program—Business Case for Power Company Participation, 2008," was shared with EPRI mem-bers in 2008, but not formally published until 2010. The estimates of when various TMDLs would be issued therefore takes 2008 as its base year, meaning that Ohio was expected to issue its standard that same year, and Kentucky in 2010. At the time of writing, in the summer of 2015, none of these states has issued its limits.

10 All CIG recipients can be found on: http://www.nrcs.usda.gov/wps/portal/nrcs/ciglanding/ national/programs/financial/cig/cigsearch/. Last accessed on 08/17/2015.

in place and several key collaborators expressing support, EPRI could begin in earnest.

Challenging currencies

Around that same time, as EPRI worked to secure what would eventually be more than $7 million to fund its market creation efforts, the two other Water Quality Trading efforts in the Ohio River basin also received Targeted Watershed Grants from the EPA. One grant funded an attempt by The Ohio State University to expand the Sugar Creek program to a larger area. The other funded the continuation of a Water Quality Trading project in the Great Miami River basin, led by the Miami Conservancy District.[11] The Great Miami River is a tributary of the Ohio River. The Miami's watershed is located in Ohio, and makes up part of the much larger Ohio River basin.

The Great Miami River Watershed Water Quality Credit Trading Program, as it is formally known, was created in 2003, with credit trades occurring every year since 2004 (Newburn and Woodward 2012). After more than a decade of buying and selling, almost 400 nutrient removal projects have been installed on farms throughout the watershed. The Ohio EPA, the Department of Natural Resources and several other agencies were closely involved in the development of this program and it continues to operate (as a pilot project) as of this writing.

Initially, it seemed logical that this program would function as a model and eventually perhaps an important part of EPRI's much larger Ohio River Basin Trading Project. An expert who had helped develop the Great Miami trading program was hired by EPRI as a consultant, and Miami Conservancy District staff attended many of the meetings between the key stakeholders.[12] In some important ways, EPRI's vision for an ESM was more ambitious and complex than those existing programs, as EPRI's hoped-for ESM included trading across state boundaries and in multiple types of ecosystem services, specifically offsets for greenhouse gas emissions (EPRI 2013a).

The third way in which EPRI's program appeared more complex than these existing programs was in the precise way in which credits would be calculated. The Sugar Creek and Great Miami River basin trading programs use a relatively simple Excel spreadsheet, formally known as the EPA Region 5 Pollutant Load Reduction Model, to calculate their nutrient credits. The

---

11 For brief descriptions of all recipients of Targeted Watershed Grants for Water Quality Trading Projects in 2008, see: http://water.epa.gov/type/watersheds/trading/twg_index.cfm. Retrieved on 8/21/2015.

12 Author interview with Miami Conservancy District Manager of Watershed Partnerships on October 3, 2015.

Indiana Department of Environmental Management created this spreadsheet based on data published by Michigan's Department of Environmental Quality in 1999. Its primary use is to estimate how many pounds of nitrogen and phosphorous are prevented from flowing into a waterway as the result of installing a specific Best Management Practice. The approach is widely used within and outside of EPA Region 5 to assess the effectiveness of investments in water quality improvement. The existing trading programs in Ohio rely on this spreadsheet as a central element in their measurement system in part because the U.S. EPA, the Ohio EPA and the local Soil and Water Conservation Districts are familiar with this method. Specific staff members in all these organizations routinely use it to determine the effectiveness of BMPs funded through government programs.

EPRI did not immediately determine how to conduct credit calculations for its market. In fact, the organization commissioned a detailed analysis of two models that were new to the region, namely the Watershed Analysis Risk Management Framework (WARMF) and the Nutrient Trading Tool (NTT) (EPRI 2011). The NTT can be used to estimate the effect of the BMP at the "edge-of-field" and would effectively take the place of the Region 5 spreadsheet. The WARMF model calculates the attenuation of the nutrient reduction between the restoration site and the water body of concern, in this case the water downstream of the power plant that is discharging nutrient-laden water (Keller et al. 2014).

WARMF could be an add-on to either the NTT or the Region 5 spreadsheet, to provide a more precise calculation of the effect of a BMP miles downstream of its location. Based on an in-depth test of the NTT and WARMF, a survey, multiple listening sessions with agricultural producers and comments from key agencies, EPRI made the choice for a combination of the Region 5 model and WARMF. This made it easy to connect to existing expertise, for example in Ohio's Department of Natural Resources and the various Soil and Water Conservation Districts.

The Miami Conservancy District did not see the choice to add the WARMF model to the credit calculation method as useful, or even desirable. The Ohio Administrative Code formally prescribes the EPA Region 5 Model[13] for all Water Quality Trading programs in the state. This is also the calculation method used in the Sugar Creek and the Great Miami River trading programs. The EPRI approach is in line with this statute, since the Region 5

---

13 Chapter 3745-5 of the Ohio Administrative Code, titled "Water Quality Trading Program," includes a description of the "Load Reduction Spreadsheet" with a link to the Region 5 Model on the Ohio Department of Natural Resources website. This part of the code is available at: http://codes.ohio.gov/oac/3745-5. Retrieved on 08/29/2015.

Model is still used. However, from the perspective of the existing trading programs in Ohio, EPRI's insistence on the need for additional modeling (namely WARMF) on top of the Region 5 Model was an implicit challenge to the precision and accuracy of its own credits.

The Miami Conservancy District and its partners had struck a balance between science, state and market by using the Region 5 spreadsheet and a system of trading ratios that require credit buyers to buy more credits based on the part of the watershed in which they are located (Newburn and Woodward 2012, 160). This kind of balance is very difficult to achieve (Robertson 2006). Yet EPRI argued that significant additional scientific work needed to be done to show the equivalence between implementing a Best Management Practice on a farm on one hand, and the polluted water power plants discharged on the other (EPRI 2014). The role of the WARMF model was specifically to account for the distance between the two locations by calculating the attenuation effects. This difference of opinion effectively ended the close collaboration between the Miami Conservancy District and EPRI, which each trading program continuing on a different path related to measurement. Both trading systems continue to operate as pilot projects, with the odd result that it is possible to purchase two very different types of credit for a pound of nitrogen in Ohio.

These early stages of EPRI's foray into creating an ESM reveal that it is possible for two different measurement systems to emerge in the same watershed, under the same regulatory regime and for the same (ecosystem) service, namely nutrient reduction. So not only do measurement systems for ESMs need to strike a delicate and often unstable balance, they also need to compete with alternative approaches to measuring the same phenomena. While both credit calculation methods continue to exist side by side, EPRI has not pursued the development of credits inside the Great Miami River basin, instead focusing elsewhere in the Ohio River basin to pilot-test its methodology.

The Steam Electric Category and the limits of equivalence

In 2011, with key partners in place and a sophisticated measurement system in development, EPRI was able to start approaching landowners to see if they were interested in implementing Best Management Practices on their farms. EPRI remained hopeful that these credits could eventually help power plants offset high nutrient levels in their effluent and thereby provide the energy companies an affordable way of meeting requirements under the Clean Water Act. Therefore, the organization focused on agricultural regions upstream of its members' facilities, especially targeting parts of the watershed where significant sources of nutrient from agricultural land use were identified. The Miami Conservancy District made clear that the organization did not want landowners in that watershed pursuing these credits, to avoid competition between the trading programs. So while the Miami River basin had long been a target area within the Ohio River watershed, EPRI focused elsewhere.

The first projects were installed in Indiana, Kentucky and Ohio under the auspices of the Ohio River Basin Trading Project in 2013. Once these BMPs were formally installed and assessed using the newly developed measurement system, three of EPRI's member organizations purchased credits. Duke Energy, American Electric Power (AEP) and Hoosier Energy collectively bought 9,000 credits. These power companies made clear they would not use these credits to meet any potential future permits. Rather, this purchase was an act of corporate sustainable responsibility.[14] The continued absence of a regulatory requirement to reduce their nutrient levels made the owners of these power plants unlikely to continue purchasing these credits on a voluntary basis.

The second round of credits, generated from projects installed on farms in Ohio, Kentucky and Indiana in 2014, was much more difficult to sell. In the spring of 2015, EPRI repeatedly announced an auction, to take place in New York City, in which these credits would be put up for sale. The revenue from those sales would be used to fund additional ecosystem services creation and restoration in the Ohio River basin. In May 2015, the auction was postponed indefinitely.[15]

Almost coinciding with the installation of the first round of credit-generating projects in 2013 was the U.S. EPA's release of its Proposed Effluent Guidelines for the Steam Electric Power Generating category. These guidelines are based on the Clean Water Act, and make it much less likely that Duke Energy, AEP and Hoosier Energy will become credit buyers anytime soon. The new rule, finalized in September 2015, replaces an existing standard, which was issued in 1982. This regulation directly applies to the power plants EPRI has been targeting as potential credit buyers. These Effluent Guidelines are technology based, meaning that they prescribe the basic characteristics of the filtration technologies that all power plants using steam to generate electricity have to install.

This new rule sets specific allowable limits on the quantity of pollutants in water discharged by power plants. Elements listed in the guidelines include arsenic and zinc, of which only trace amounts are allowed in effluent based on the new rule. Broadly speaking, these standards are based on the EPA's

---

14 See EPRI's Spring 2015 newsletter for a description of this purchase. Available at: http://wqt.epri. com/pdf/Ohio%20River%20Basin%20March%202015r1.pdf. Retrieved on 08/29/2015.

15 For one of the announcements, see: http://wqt.epri.com/pdf/Ohio%20River%20Basin%20 March%202015r1.pdf. Retrieved on 07/03/2015. The cancellation of the auction was announced on EPRI's Web portal on Water Quality Trading: wqt.epri.com. Last accessed on 09/04/2015.

assessment of what the current best available technology is for the specific type of effluent-producing activity, and therefore the guidelines need to be updated with some regularity to keep up with advances in filtration technology, but also to deal with new industrial practices. For example, besides the Steam Electric category, the EPA is also working on new guidelines for unconventional extraction in the oil and gas industry and for certain dental practices. The basic idea is that these technology-based limitations provide a minimum quality standard that every facility in a specific category needs to abide by when it discharges water. In areas where water quality is already poor, additional water-quality based standards can be issued through the TMDL process.

Under the EPA's 2003 Water Quality Trading policy, technology-based standards cannot be met through trading. So by including strict limits on nitrate as a pollutant that facilities in the Steam Electric Category need to meet under the technology-based guidelines, the U.S. EPA essentially removed the central regulatory incentive for power plants to buy EPRI's water quality credits. In EPRI's comments on the EPA's draft of these guidelines, the organization states:

> However, because the EPA WQT Policy does not support trading to comply with a technology-based effluent limit (TBEL), in the absence of specific authorization, inclusion of technology-based effluent limitations for nitrate/nitrite in the ELG [Effluent Limits Guideline, TVM] will eliminate any reasonable opportunity for using WQT as a cost-effective compliance tool, and will also eliminate associated ancillary benefits (pollinator habitat, carbon sequestration, social benefits to farmers, improved agricultural practices, improved soil health, etc.). (EPRI 2013b, 37)

Simply put, these guidelines require all the power plants that EPRI viewed as potential credit buyers to upgrade their on-site filtration processes to remove nitrogen from their effluent. In its comments on the draft guidelines, EPRI suggested that the EPA include a specific authorization to allow dischargers in the Steam Electric category to meet the technology-based effluent limit for nitrogen through Water Quality Trading, but the U.S. EPA did not include this exception in its final guidelines.[16] Many environmental groups strongly oppose these exceptions on principle.

The inclusion of nitrates in the Technology-Based Effluent Limitations Guidelines is logical when looking at the EPA's approach to implementing technology-based standards under the Clean Water Act. The EPA categorizes

---

16 See: http://www2.epa.gov/eg/steam-electric-power-generating-effluent-guidelines-2015-final-rule. Last accessed on 10/07/2015.

industrial activities that discharge polluted water into the nation's streams and rivers, and reviews the range of elements typically found in that effluent. Based on an assessment of what technologies are available to the industrial sector, the EPA mandates the best available technology and sets specific limits on all the relevant pollutants.

By regarding Steam Electric Power Generating facilities as a single category of point-source pollution, it becomes very difficult to argue that the installation of agricultural Best Management Practices are meaningfully equivalent to the impacts of power plants. The barrier this new set of guidelines erects to EPRI's attempt to create an ESM in the Ohio River basin is significant. Much of the organization's efforts have focused on producing a measurement system that allows for the calculation of the equivalent quantity of nitrogen and phosphorous when comparing the point of discharge and the location of the BMP. However, the EPA's guidelines challenge a much more basic type of equivalence, namely that between a complex activity like electricity generation, which has a broad range of impacts on water quality, and the installation of new practices on farms, which might have a whole host of consequences as well, but of a very different type and quality.

The smaller Water Quality Trading programs in Ohio are not likely to be affected by the EPA's new rule, since their credit buyers are respectively a cheese manufacturer (in the Sugar Creek watershed) and sewage treatment facilities (in the Great Miami River basin), neither of which falls under the Steam Electric category. Both programs do continue in the pilot phase, since the Ohio EPA has not set the numeric nutrient criteria for the Ohio River basin under the Water Quality Based Effluent Limits.

EPRI's expectation that these limits would be set in 2010 in Ohio has turned out to be off the mark by at least five years. So while EPRI started out by selecting the Ohio River basin as the site for its "demonstration project" in part on the grounds that a regulatory driver for credit trading was imminent, almost 10 years later that requirement has not materialized. In fact, for the category of water dischargers that EPRI is most interested in, those criteria have become irrelevant as a result of the new Steam Electric Effluent Guidelines. So while EPRI has been able to count on monetary support from the most relevant federal agencies, namely the EPA and the USDA, each of which supported the creation of the ESM-infrastructure with significant grants, the formal regulatory process has proven more difficult for EPRI to influence.

The challenge of producing equivalence between power generation and farming improvements shows that the most basic assumptions on which measurement systems for ESMs are founded can easily become contested. Even when a sophisticated method to calculate certain elements of a potential credit generating restoration project exists, the specific requirements of ESMs necessitate active support from a wide range of people and organizations, including regulators, environmental advocates, credit buyers and credit sellers. Resistance, even when passive, can become problematic enough to prevent widespread use or acceptance of any measurement system, and therefore prohibit the success of the entire ESM.

3.c No Two Wetlands Are the Same

In Oregon, the Willamette Partnership worked on four types of environmental trading regimes during the Counting on the Environment (COTE) process described in Chapter 2. The organization hoped to integrate these four credit types, for shade, salmon habitat, wetlands and upland prairie habitat, into a single ESM: the Willamette Marketplace (see Table 3.1). The measurement systems for the two landscape-based credit types, namely wetlands and upland prairies, were designed to calculate a range of ecosystem services produced on a site of either type. This is a very different approach than systems like the Shade-a-Lator or EPRI's combination of WARMF and the EPA Region 5 spreadsheet. The latter two measurement systems were developed to credit individual ecosystem services, like water cooling or nutrient removal.

The Oregon Rapid Wetland Assessment Protocol (ORWAP) and the Prairie Habitat Calculator, as they are known, are different from the models and calculators in the Ohio River basin, although they do face many of the same challenges. The intent of landscape-based ecosystem services measurement is to establish equivalence between locations where wetlands or prairies have been damaged or removed and locations where wetlands or prairies have been restored or created. The production of equivalence appears to be a much more straightforward affair in this context, since the character of the activities that need to be connected to each other is very similar. The differences between destroying a wetland and restoring a different wetland seem easier to calculate than the differences between power generation and farming.

Table 3.1 Credit types and measurement systems during Counting on the Environment

Credit Type	Measurement System	Creator
Prairie Habitat	Upland Prairie Calculator	Paul Adamus
Salmon Habitat	Ecometrix	Parametrix
Water Quality—Temperature	Shade-a-Lator	OR Department of Environmental Quality
Wetlands	OR Wetlands Assessment Protocol (ORWAP)	Paul Adamus

Source: Willamette Partnership, 2009.

During the COTE process and since then, much work has been done to create, adjust and improve these measurement systems. The outcomes of this work again show that the easy-to-use credit measurement tools ESM creators envisioned face considerable resistance in practice, often from unexpected parties. Ultimately, despite extensive negotiations and the continued evolution of elaborate credit calculation procedures, both relevant government agencies and experienced restoration professionals in Oregon remained ambivalent about ORWAP and the Prairie Habitat method. In the case of ORWAP, implementation has been limited to state-funded projects. After many years of work on this measurement system, there is substantial uncertainty about its future. The Prairie Habitat Calculator has not been used at all to develop credits within the context of an ESM or any kind of similar institution. The explanation of this lack of uptake, despite years and hundreds of thousands of dollars of investments, requires some explanation of technical details, which will be provided in the next section.

Searching for equivalence in wetland mitigation banking

Oregon has allowed *wetland mitigation banking*, a practice to protect and restore wetlands under the Clean Water Act and Oregon's Removal-Fill program,[17] for many years. In general, anyone who fills, drains or otherwise degrades a wetland is required to compensate for the damage they cause if it cannot be avoided or minimized. Compensation can be achieved by restoring wetlands in some other location or by simply buying *wetland credits* from someone else who has already restored or built a wetland. Wetland banking, as it is sometimes called, is a complicated field[18] with a long history.

Rather than trying to create a completely new approach, the participants in the COTE process decided to build on a measurement system already under development. They selected the Oregon Rapid Wetland Assessment Protocol (ORWAP), and the participants in the COTE process intended to use it to calculate the "ecosystem services production" of wetlands. They hoped to ensure the precise equivalence of a wetland destroyed and one restored as a means of compensation. ORWAP was originally developed at the request of the Oregon Department of State Lands. This agency oversees Oregon's wetland mitigation banking program. Given the long history of wetland banking, ORWAP was not the first or only effort to ensure the equivalence between impact and restoration in Oregon.

---

17 See: http://www.oregon.gov/DSL/PERMITS/Pages/r-fintro.aspx. Last accessed on 07/16/2015.
18 For a description of the complicated history of wetlands regulation, definition and classification, see: Lewis, 2001.

The "hydrogeomorphic" (HGM) method predates ORWAP and remains in use in Oregon and much of the rest of the United States today.[19] It is based on a characterization of *reference standard* sites. It highlights any deviation from those ideal-typical conditions at a potential restoration site. The HGM method identifies five basic types of wetlands along with local and regional variations. Establishing locally specific reference standards and comparing particular wetlands across these types and local standards is difficult. In practice, a ratio is often applied to determine how many acres of wetlands need to be restored. For every acre of degraded wetland, three acres have to be restored. This ratio-based approach to mitigation is common throughout the United States, but it also allows low-quality wetlands, from an ecological point of view, to be used to compensate for the destruction of highly valued ones. This has resulted in a series of legal challenges[20] as well as an extensive study by the National Research Council (2001). That study, conducted by a panel of renowned experts in the field of wetland ecology, revealed that many of the restored wetlands were poorly maintained and of limited ecological value.

These legal challenges, the National Research Council's broad critique of wetland restoration and the fact that reference standard sites are only available for a limited number of wetland types and locations in Oregon, motivated the development of a more comprehensive and precise measurement system for wetlands in Oregon. The Department of State Lands received funds from the EPA to work on ways of assessing all five types of wetlands, facilitating quantitative comparisons throughout Oregon.

The perceived low environmental quality of compensatory wetlands continues to motivate efforts to improve the environmental effectiveness of wetland mitigation banking. The creation of ORWAP is an example of one such effort, in both environmental and legal terms. Wetland mitigation banking predates the emergence of the ecosystem service concept.[21] Despite this sequence, an important part of the literature on ESMs in the United States

---

19 For an introduction to the HGM approach, see: http://el.erdc.usace.army.mil/wetlands/overview. html. Last accessed on 07/17/2015.

20 A useful overview of the history of litigation that is relevant to wetlands can be found on Prof. Morgan Robertson's personal website, morgan-robertson.com. Important cases with direct implications on the determination of what is and is not a "jurisdictional" wetland include: *United States v. Ashland Oil* (1974), *NRDC v. Callaway* (1975), *United States v. Riverside Bayview Homes* (1985). Together, these and other cases emphasized that the Clean Water Act applies to waters that are "navigable" or "adjacent" to "navigable" waters. Navigable and adjacent wetlands are protected by the Clean Water Act and fall under the purview of the federal government as a result of the Commerce Clause. This means that establishing a hydrological connection between wetlands and waterways, as well as physical proximity, is an important part of the measurement systems used in wetland protection and banking.

21 The first wetland mitigation bank, the Tenneco LaTerre Mitigation Bank in Louisiana, was established in 1984. For two very different timelines of the history of mitigation banking in the

refers to wetland mitigation banking as an important example of an ESM (Shabman, Stephenson and Shobe 2002; Womble and Doyle 2012).

The development of and later adjustments to ORWAP highlight the challenges to creating measurement systems for ESMs. The original goal of this metric was to produce a detailed and reliable accounting of the functions of a wetland, both before and after any adverse impact, and especially before and after restoration. The calculation of the ecosystem services provided by a wetland was added later in an attempt to simplify the outcome of the measurement system and produce a single numerical score. In other words, ORWAP is an attempt to connect the decades-long history of wetland mitigation banking to the requirements of ESMs. In addition, ORWAP significantly influenced the development of measurement tools in other emerging ecosystem services markets.

Nevertheless, an analysis of the current wetland credits available for sale in Oregon shows that, out of 22 wetland restoration "banks," only three have used ORWAP's ecosystem services–based method[22] to calculate credits. Those three have all been undertaken by the State of Oregon, meaning that the uptake of this new measurement system among private and municipal wetland bankers has been nonexistent.

Quantification and simplification: The ORWAP spreadsheet

At the center of ORWAP is a series of functions, coded into an Excel spreadsheet. To count the ecosystem services a particular wetland provides, ORWAP users rely on aerial photography to delineate the wetland. Then they gather additional data using a variety of online tools (like the soil type using the NRCS soil survey). Finally they fill out ORWAP "data forms" (Adamus, Morlan and Verble 2009a and 2009b). These forms are essentially long lists of questions about the wetland, for example, what kind(s), if any, of non-native aquatic animals can be found (F31), which herbaceous species are dominant (F42) and whether the timing of water flow into the wetland has changed recently (S5). Once those data forms, which include 140 such indicators, are completed, the Excel spreadsheet calculates a set of outcomes.[23]

United States, see: http://www.msusa.com/timeline/ (developed by a mitigation banking firm) and http://blog.as.uky.edu/robertson/?page_id=197 (developed by Morgan Robertson). Both last accessed on 07/15/2015.

22 See: http://www.oregon.gov/dsl/PERMITS/Pages/mitbank_status.aspx. Last accessed on 07/15/2015.

23 Each ecosystem function or value score is calculated based on the relevant indicators, using a specific "scoring model." For example, the Pollinator Habitat score is estimated as the average of three groups of indicators, namely (1) the average of: percent of surrounding area comprised of native vegetation, proximity to closest patch of native vegetation and size of that patch; (2) the average of: percent cover of native (vs. non-native) herbaceous plants and especially

These calculations include a whole host of assumptions and simplifications which have very little to do with "pristine" wetlands. In fact, the scoring is more like a composite report card grade than an actual measure of the disturbance a wetland has experienced. Once the questions have been answered, the spreadsheet calculates 16 *functional effectiveness scores*, categorizing the outcomes into functions like *Organic Matter Export* and *Waterbird Nesting Habitat*. These scores are then aggregated into five *grouped functions*, the average of which produces the eventual outcome, in a unit defined as *functional acreage*. This unit can range between zero and one, based on the idea that one acre of "pristine" wetland hypothetically produces a score of one. At the same time, one acre of completely degraded or developed wetland would produce zero functional acres. The value of calculating this composite score for market purposes is to conduct an analysis before and after an intervention on a wetland. Be it restoration or degradation, the difference between the pre- and post-action scores produces the number of credits to be bought or sold, when multiplied by the total acreage of the site.

This process makes the overall calculation of *functional acreage* somewhat analogous to a wetland's grade point average (GPA). The ORWAP manual clearly states that the individual function and value scores are more like a report card score than a straightforward calculation of wetlands' ecological functioning.[24] A wetland biologist with the Oregon Department of State Lands, the agency overseeing Oregon's statewide wetland mitigation banking program, explained that trying to move from the conventional, more narrative description of wetlands to this ecosystem services–based measurement system was not an easy decision: "The thing we were stuck on the most was how do you compare—how do you get your final number of credits" (Interview with

non-graminoids, percent cover of native (vs. non-native) woody plants, extent of downed wood, extent of snags and large-diameter trees and presence of regionally uncommon herbaceous plants; and (3) the average of: percent of wetland not submerged by persistent water, infrequent fires and vegetation removal, vegetation height diversity, intermediate ground cover, large extent of microtopographic variation, presence of nearby rocky areas and minimal soil disturbance. In slightly more mathematical terms, this is expressed as: (POL) = [AVERAGE(natveg,natvprox,na tvacres) + AVERAGE(gramin,herbsens,herbrare,woodynn,woodydbh,downwood) + AVERAGE (lomarsh,persist,firehay,htunif,gcover,girreg,cliff,soildisturb)] /3

All the scoring models are described in these narrative and mathematical terms in Appendix B of the ORWAP Manual, along with the results of the repeatability and sensitivity analyses. The Manual is available at: http://www.oregon.gov/dsl/WETLAND/docs/orwap_manual_v2.pdf. Last accessed on 07/11/2015.

24 As is stated on page 4 of the manual: "The numeric estimates ORWAP provides of wetland functions, values, and other attributes are not actual direct measures of those attributes, nor the products of validated mechanistic models of ecosystem processes. Rather, they are estimates of those attributes arrived at by using standardized scoring models that systematically combine well-accepted indicators" (Adamus, Morlan and Verble 2009b, 4).

author, July 18, 2011). The reason it was so difficult was because the members of the working group considered a specific scenario undesirable:

> So if we go to a function-based system and somebody has to get—has to replace two acres of fairly high-quality habitat, they can get there by doing 20 acres of kind of low-quality replacement, and that's the issue that we've been—that caused the most discussion, I think, and the most heartburn over how it's being done.

The negotiated solution that resolved this dispute during the COTE process was based on two principles. The first was a series of additional ratios for *priority areas*. This reflects that even highly elaborate measurement systems cannot readily solve the problem of effectively protecting special places into the design of ESMs. In fact, even a sophisticated and "robust" measurement system like ORWAP was still not trusted to accurately take into account the ecological characteristics that could make a certain wetland simply too special to allow its degradation.

The second principle the participants adopted to counter the limited ability of their measurement system to prevent the destruction of the most special wetlands was a final bureaucratic check, described by one of the participants as follows:

> So what we decided was a peer-functioned scoring system doesn't take the review of the mitigation plan and the ability to say yes or no away from the agency, it's just another step that we have to do and evaluate. (Interview with author, July 18, 2011)

While this appears to be a straightforward reassertion of bureaucratic discretion, the participants realized the complexity of adding a further decision-making step:

> But it's hard then to argue with, say, a mitigation banker that comes in with a proposal and they want to apply the math and get this many acres—this many credits, you know, for their mitigation bank, and you have to do that before you really know what the impacts are going to be.

In this regulator's view, an important reason this additional check on mitigation banker ambition is discomforting is the fact that a professional mitigation banker can do the math. This would place the regulator in a situation where she would have to argue with developers and mitigation bankers about the equivalence between wetland impact and wetland restoration, notwithstanding the existence of ORWAP.

If the regulator decides to deny a wetland permit based on this broad review of the mitigation plan, she has to argue that the ORWAP score calculated by a mitigation banker indicates "equivalence" between restoration activities and impacts. She knows that such equivalence has not conclusively been established. In a context where rationality, science and precision are highly valued, there is serious discomfort about relying on a market framework—supply and demand of supposedly fungible credits—to determine the location and acceptability of wetland restoration projects.

Despite all of this work, and the general agreement reached on this calculation method, some of the professionals involved in the creation of wetland restoration sites are not particularly excited about ORWAP. The scores before and after the restoration tend not to be very far apart. Basically, when using ORWAP to calculate the credits for a specific restoration project, the outcomes reflect very little increase in the production of ecosystem services by that wetland. To stick with the report card metaphor, it's hard to go from a C to an A in ORWAP. Most of the scores are clustered around a B-. The natural resources manager for the City of Eugene, which runs its own mitigation bank, describes this problem as: "One of the things I've learned from the ORWAP methodology is that it's very insensitive to restoration efforts" (Interview with author, August 18, 2011). Eugene built a new, award-winning wetland bank after the ORWAP methodology had been developed, but decided to calculate credits[25] under the old calculation method. In the end, for this credit developer, a simple comparison between ORWAP and the older, simpler and ratio-based so-called HGM approach was all that was needed: "And so we ran both the HGM methodology and the ORWAP to see, what did each one say in terms of how many credits we would generate, and the HGM approach came out generating way more credits than the ORWAP" (Interview with author, August 18, 2011).

The low credit estimates produced by the measurement system, combined with the regulator's ambivalence about letting its output function as the primary means of making permitting decisions, have resulted in a situation where the mitigation bankers are understandably reluctant to invest significantly in its use. In the case of ORWAP, there was an existing credit determination system, based on ratios, already in place. This outcome is somewhat analogous to EPRI's experience in the Ohio River basin, where an existing and less sophisticated approach to evaluating an environmental phenomenon turns out to be difficult to unseat. While the basic observation that engrained practices are hard to change is commonplace, its relevance to measurement systems for ESMs is hard to overstate. The most ambitious efforts to create markets for

25 See: http://www.oregon.gov/dsl/Pages/pr1209_slb_awards.aspx. Last accessed on 07/15/2015.

ecosystem services in the United States all rely on or at least anticipate some form of a regulatory driver. Yet the policies these regulations are based on are decades old, meaning measurement and evaluation practices are typically either well established or have been at the center of acrimonious debates for many years. The introduction of a concept like ecosystem services does not fundamentally change the interests of those actors who have been active in the relevant environmental policy realm.

A fluttering market

The Upland Prairie Calculator is the second landscape-based measurement system developed during the COTE process, and was modeled after ORWAP. It is also a spreadsheet-based calculation of *functional acreage*, developed primarily by the same ecologist responsible for ORWAP.[26] The experience of tailoring ORWAP to increase its ease of use for wetland banking directly informed the development of the Prairie Calculator. Like ORWAP, the Prairie Calculator was regarded as a way to further develop a "banking" practice in Oregon. But federal law does not explicitly protect prairie as a type of landscape in the way the Clean Water Act protects wetlands. Rather, the Prairie Habitat Calculator was designed to improve and popularize "conservation banking," also known as "species habitat banking" or "biodiversity banking."

The creation of the Prairie Habitat Calculator is directly and deeply informed by the ecosystem services discourse and the goal of creating a market. While the calculator has not been used to facilitate actual transactions, it does provide a window into the constraints and opportunities associated with developing ecosystem service measurement tools for markets. Specifically, the story of the Prairie Habitat calculator shows that an emphasis on reducing transaction costs creates pressure to reduce the complexity of the measurement system involved. The goal is to make implementation possible with only a brief visit to the restoration site.

From the point of view of market proponents, the Endangered Species Act (ESA) operates much like the Clean Water Act. Once a particular species is listed as endangered, the ESA can be used to require compensatory actions to offset adverse impacts on the habitat of that species. The basic logic of conservation banking mimics wetland banking. A developer who destroys an environmentally sensitive area can be required to compensate for that destruction

---

26 Unlike ORWAP, the Upland Prairie Habitat Calculator was not formally published, and primarily exists online on the Willamette Partnership's website, together with a brief manual for its users. Both can be viewed here: http://willamettepartnership.org/wp-content/uploads/2015/02/Upland-Prairie-Habitat-Quality-Calculator-v2.0_USER-GUIDE.pdf. Last accessed on 07/03/2015.

by restoring or conserving (in perpetuity) a landscape elsewhere. However, important differences exist between the Endangered Species and Clean Water Acts. The Endangered Species Act is focused on the preservation of populations of a species, which can often move around, while the Clean Water Act, at least as it relates to wetlands, is primarily about the protection of geographically defined places.

Conservation banking emerged in the 1990s, around the same time as wetland banking became of more interest to private entities and for-profit restoration projects.[27] However, this type of environmental banking is much less common than wetland banking. As described in Chapter 2, this type of environmental trading is primarily used in California,[28] where conservation banking has become fairly routine. In 2011, there were 89 active conservation banks in California, compared to 20 in the rest of the United States.[29] Only one of those 20 conservation banks was in Oregon. Given the variety of endangered species to which conservation banks apply, a great many credit types are involved, both for habitats and individual species.

In the case of wetlands, 40 years of jurisprudence and scientific efforts to define and categorize them informed the creation of ORWAP. In addition, the five basic categories of wetland have become relatively well established during that period. By contrast, the listing of individual species as endangered drives species habitat banking. Some of the species involved are relatively obscure, like the Fender's blue butterfly. Only a handful of people are dedicated to studying them full time. Species habitat banking has also not been the subject of the kind of litigation that wetland banking has. This means that questions about the exact definition of how to establish jurisdiction or define the geographic extent of a habitat have not been settled in the kind of public, detailed and formal way that the Supreme Court has for wetlands.

In Oregon, the participants in COTE wanted to increase the feasibility of species habitat banking by developing the Prairie Habitat metric and the associated rules for generating credits. As the executive director of the Willamette

---

27 A useful source of up-to-date information on conservation banking is the website: www.species-banking.com. The State of California issued guidance on conservation banking in 1995, available at: http://ceres.ca.gov/wetlands/policies/mitbank.html. Following the growth of conservation banking, the U.S. Fish and Wildlife Service issued national guidelines in May 2003. This is available at: http://moderncms.ecosystemmarketplace.com/repository/moderncms_documents/Federal%20Guidance%20on%20Conservation%20Banking%202003.pdf. Last accessed on 06/03/2015.

28 For the most up-to-date official information on the type and number of wetland and conservation mitigation banks in the United States, the U.S. Army Corps of Engineers' Regulatory In-lieu Fee and Bank Information System (RIBITS) is available at: http://geo.usace.army.mil/ribits/index.html. Last accessed on 07/28/2015.

29 See note 10.

Partnership, at that time still working for Clean Water Services, describes the composition of the prairie working group: "We wanted to make sure U.S. Fish and Wildlife Service and the Oregon Department of Fish and Wildlife were there. They had to say yes" (Interview with author, July 8, 2011). A different member of that working group, an employee of the City of Eugene, describes the relationship between the development of a measurement system and its potential policy implication as follows:

> We do have an endangered butterfly. It's called the Fender's blue butterfly, it occurs in upland prairies. And then there's its host plant, the Kincaid's lupine, which is also listed under the Endangered Species Act. And so there was some discussion of okay, there's not a market for it yet. You're trying to create a new mechanism or a new market for it and at the same time there is an avenue via wetland banking. (Interview with author, August 18, 2011)

This city employee from Eugene is referring to the possibility of incorporating the presence of the Fender's blue and the Kincaid's lupine into ORWAP, by giving "extra points" for protecting those species in a wetland. In fact, both species are represented in ORWAP's supplemental datasheet,[30] showing the difficulty of maintaining hard and fast distinctions between wetlands and uplands.

The Prairie Habitat Calculator was eventually developed to assess the "functionality" of a prairie related to the habitat requirements of several listed species, but mainly focused on the Fender's blue butterfly (*Icaricia icarioides fender*). This butterfly is found only in certain parts of the Willamette Valley. It was listed as endangered in 2000.[31] Creating a useful calculator was something of an experiment since this particular type of landscape has not received the level of scientific, legal and political scrutiny wetlands have. The wetland scientist responsible for ORWAP was also in charge of the work on the Prairie Calculator. This is not his area of expertise, and he quickly determined that it would be difficult to model the Prairie Calculator after ORWAP:

> The first thing we decided was that there was not enough science to parse out the individual ecosystem services of upland prairies but rather what we thought was a more realistic objective was to just have one overall score for an upland prairie and that score would be based on things that

---

30 Kincaid's lupine can be under the nitrogen-fixing plants, row 50 and the Fender's blue under rare invertebrates, row 46. This sheet can be downloaded at: http://www.oregon.gov/DSL/WET-LAND/Pages/or_wet_prot.aspx. Last accessed on 07/10/2015.

31 For more information on the Fender's blue butterfly and its listing, see: http://www.fws.gov/oregonfwo/Species/Data/FendersBlueButterfly/. Last accessed on 07/03/2015.

we thought would kind of generally indicate upland prairies that are most functional and of highest value. (Interview with author, July 15, 2011)

This led to a far simpler metric as compared to ORWAP. About 20 indicators have to be filled out in the Upland Prairie Calculator spreadsheet. The final score is only expressed in functional acreage, meaning that there are no specific scores for particular ecosystem functions or services. Some participants regard this simplicity as an important benefit, since it means that field visits can be relatively short. Other participants believe that certain important indicators, like the presence of housing near the site, should have been incorporated.[32]

Much of the discussion in the working group that developed and reviewed the relevant metrics concerned the preservation of existing prairie and encouraging the restoration of landscapes so that they functioned as effective prairie habitats. As one of the participants describes these conversations:

There was a desire to score highly the highest-quality upland prairie sites. It became more of a discussion of, well, shouldn't—what should this metric be measuring? Should it be measuring the amount of uplift you could get from restoring a degraded site or do we want to actually weight this whole thing so that protecting those few remaining high-quality sites gets scored high? And so it might not actually be geared towards restoring like the wetland banking world is, but rather conserving. (Official from the City of Eugene, interview with author, August 15, 2011)

Several indicators can easily be recognized as examples of an emphasis on preservation. The presence of non-native herbaceous vegetation (Question 8 in the calculator) and the extent to which soil has been disturbed through compaction or plowing (Question 11) are both worth more "points," potentially, than the current land management regime (Question 10). The latter two might present an easier opportunity to render the prairie more habitable for particular species like the Fender's blue butterfly, but that was not the priority of the participants when the calculator was created.

Unlike ORWAP, the Upland Prairie Calculator is not based on three decades of regulatory and scientific work. Thus, it is much more likely for prairie experts to find elements of the measurement system with which to disagree. Essentially, after the COTE process, there was no further organized scrutiny of this metric, by experts, stakeholders or the public. Yet the Prairie Habitat metric did not completely disappear when the COTE process was over. The

---

32 Official with the Oregon Department of Land Conservation and Development, interview with author, July 18, 2011.

measurement system became something of a model for a much larger effort, at least in terms of geographic scale and overall ambition, to measure other parts of ecosystems.[33] Given the fact that the Prairie metric was never used for its intended purpose, this is somewhat surprising.

The fact that little disagreement has emerged over the Prairie Habitat metric can be attributed to the fact that it has hardly been used in a formal trading context, in Oregon or anywhere else. Species habitat banking, for the Fender's blue butterfly or any other endangered species, remains an exceedingly rare practice in Oregon. The U.S. Department of Agriculture, through its Office of Environmental Markets, funded a review of existing measurement systems for conservation banking (and similar practices outside of the United States). The Willamette Partnership produced this report, titled *Measuring Up: Synchronizing Biodiversity Measurement Systems for Markets and Other Incentive Programs* (Cochran and Robinson Maness 2011). They found only three successful endangered species mitigation stories outside of California, in Georgia, Florida and Alabama.

The efforts during Counting on the Environment to produce measurement systems that can convincingly produce equivalencies between similar types of landscapes has turned out to be exceedingly complicated and ultimately limited in its success. Both for wetlands, with a long history of regulatory and scientific attention and for prairie habitat, a landscape type that has become a cause for concern more recently, the Willamette Partnership and its collaborators produced working measurement systems. Yet their use in ESMs has been largely nonexistent and the unwillingness to do so can be traced to key participants' concerns over equivalence, a basic requirement underpinning the very notion of an ESM.

3.d Markets and Nonequivalence

Many proponents describe the emergence of standardized measurement and elaborate accounting systems as a key benefit of ESMs and more broadly as a necessary step to successful environmental conservation (Whitworth 2015). Practical experience in the Willamette River and Ohio River basins suggests that such systems are difficult and costly to create, and that there are significant barriers to their use. The basic premise of all ESMs is that it is possible

---

33 This effort is titled "The Conservation Registry" and managed by the Defenders of Wildlife, an organization actively involved in the Counting on the Environment process. For more information, see: http://www.conservationregistry.org. Last accessed on 10/07/2015. For a description of the connection between the metric development during Counting on the Environment and the Conservation Registry, see: https://conference.ifas.ufl.edu/aces12/presentations/4%20Thursday/A-B/Session%207E/YES/1020%20Vickerman.pdf. Retrieved on: 10/07/2015.

to establish credible equivalencies, between different places, such as two wet-lands, or different environmental impacts, like water discharge from a power plant and water filtration by a riparian buffer. It is only after some measure-ment system has shown that these places or impacts are *similar enough*, that trading to achieve efficiency, innovation and cost-reduction is possible.

This chapter has shown that the development of elaborate measurement systems in conjunction with ESMs in the United States has been significant, difficult and ultimately not very successful. While the national financial crisis hurt many sectors in the United States during the same period, the develop-ment of these measurement systems was funded through USDA and EPA grants. Well-known academic experts and nonprofit organizations succeeded in creating sophisticated tools and methods to assess a range of elements and processes. Yet very few, if any, have become widely accepted or relied on to actually facilitate credit trading in an active market.

Efforts to measure credits and promote ESMs are likely to go on for some time. Organizations like EPRI and the Willamette Partnership continue to stimulate the measurement systems' acceptance, although in some situations market proponents have simply started over and are trying to develop com-pletely new metrics. As long as foundations and government agencies are willing to support the development of measurement systems, organizations will probably create more tools and techniques for ESMs. All the hard work done over the past decade in these three watersheds to produce measurement systems to facilitate trading of multiple types of ecosystem service credits, however, has not produced a lot of great examples. Given that much of the promise of ESMs has been touted as their ability to foster trading of multiple ecosystem services, and to capture the complexity of the interactions between different services, the fact that the Shade-a-Lator continues to stand out as a particularly successful example, yet captures a single aspect of environmental quality, seems to undermine one of the oft-repeated goals of ESMs.

Chapter 4

DEVELOPING PARTICIPATION

The Northwest Environmental Defense Center, the Northwest Environmental Advocates and the Idaho Conservation League filed an open records request to gather more information on what they called "secret meetings to develop regional trading policies"[1] in the spring of 2013. The Willamette Partnership organized the meetings in question to create a "Joint Regional Water Quality Trading Agreement" between Idaho, Oregon and Washington. A $1.5 million contribution from the USDA's Conservation Innovation Grant program funded the process.[2] One of the groups that filed the request, Northwest Environmental Advocates (NWEA), indicated that it was not against all forms of Water Quality Trading, but believed that Oregon's shade-trading program violated EPA regulations.[3] NWEA had long been highly critical of Oregon's water quality policies, and has successfully challenged elements of those policies in court.

This type of resistance to "closed-door" meetings between proponents and regulators highlights the third major challenge facing markets for ecosystem services. The efforts in the Pacific Northwest, the Chesapeake Bay and the Midwest show that displacing certain environmental qualities and producing equivalence between pollution and restoration have proven formidable barriers to the full realization of the ESM idea. The third type of obstacle for the creation of ESMs is the development of effective participation procedures in the creation and implementation of this new type of institution. NWEA's suspicions about the regional discussions and its strong opposition to the shade-trading program gives some insight into how difficult it can be to reach consensus on the creation of an ESM. However, the environmental advocacy organization's previous success in the courtroom and its ongoing scrutiny of

---

1 See: http://northwestenvironmentaladvocates.org/water-quality-trading-innovation-or-hoax/. Last accessed on 08/20/2015.

2 For the Willamette Partnership's press release, see: http://willamettepartnership.org/news-and-publications/willamette-partnership-receives-1-5-million-cig-grant-from-usda/. Retrieved on 08/20/2013.

3 See for example: http://www.northwestenvironmentaladvocates.org/nweafiles/2013/04/Order-on-NPS-and-ESA-Remedies_201301.pdf. Retrieved on: 08/20/2015.

market-creating efforts suggest that ignoring them altogether would also be a mistake from the perspective of the proponents of ESMs.

There is no formal handbook indicating how to structure a participatory decision-making process for the creation of an ESM at the watershed scale, although numerous prominent market proponents have written extensively about their experiences and insights (Jones et al. 2006; Kieser and Logue 2015; Willamette Partnership, Pinchot Institute for Conservation and World Resources Institute 2012; Willamette Partnership, World Resources Institute, and the National Network on Water Quality Trading 2015). While these emerging practical lessons are relevant, there also remains no clear agreement among scholars about what kind of organization should take the lead in trying to start such a process, or how to go about developing appropriate leadership (Engel, Pagiola and Wunder 2008; Johnson, White and Perrot-Maitre 2000; Salzman 2005). However, the broader literature on the role of participation in complex policy and planning processes does offer a number of suggestions and examples (Innes and Booher 2010; Susskind, McKearnan and Thomas-Larmer 1999; Wondolleck and Yaffee 2000).

This chapter presents an analysis of three participatory decision-making processes connected to the creation of ESMs. While market proponents in these three cases paid careful attention to the design and implementation of their public engagement strategies, all of them faced opposition and charges of exclusion and steamrolling at some point along the way. Given the absence of clear best practices regarding participation and the creation of these complex institutions in the "twilight zone" between public and private (Seidman 1998), the first part of this chapter will outline the framework for this analysis. This follows the work of political scientist Archon Fung and centers on three returning issues in participatory process design, namely 1) participant selection, 2) communication among the stakeholders and 3) who should have final authority.

The empirical cases are three participatory decision-making processes about proposed ESMs. The first examines the efforts to create a single water quality market for the entire watershed in the Chesapeake Bay states. This process is driven by the U.S. EPA and various state agencies, and most closely follows the participatory practices commonly associated with regulatory implementation, meaning an emphasis on public comments on draft documents and public hearings. This process has run into problems associated with the sheer scale of the proposed ESM. The key market proponents have found it difficult to determine who should be involved at which stage of the negotiation.

The second case pertains to the Ohio River Basin Trading Project. While EPRI, as a nongovernmental entity, has been able to be quite flexible and strategic in its outreach and engagement, questions of representation and participation have complicated its efforts as well. What the appropriate role for particular government agencies with historically close ties to the agricultural sector should be in the design and implementation of an ESM became a controversial issue. How to deal with groups that resisted certain aspects of the proposed ESM was another persistent obstacle in EPRI's efforts to develop a broad consensus in the Ohio River basin.

The third case, the Willamette Partnership's project to integrate four different ecosystem services trading schemes into a single ESM, has most closely followed consensus-oriented suggestions regarding process design. By bringing in an external mediator and conducting an extensive stakeholder assessment, the proponents of an ESM in Oregon invested significantly in participation. Even this level of commitment to inclusivity and deliberation did not result in the hoped for functioning, integrated ESM.

The conclusion to this chapter returns to the question of why it has been so difficult to create markets for ecosystem services in the United States. This final section highlights the specific elements of the ESM idea that make meaningful participation in its creation complicated, focusing on questions about the need for and type of ongoing public scrutiny of credit-generating environmental restoration projects.

4.a Selection, Interaction and Authority

The challenge of participation is connected to the ESM design and measurement problems analyzed in the previous chapters. For example, determining which towns, agencies or groups to bring to the table during the design process is clearly connected to the question of delimiting the geographic regions in which markets will be located, and the selection of specific priority areas. The complexity of the measurement systems influences the type and level of expertise participants will need in order to communicate effectively about ESM design.

The examples in this chapter also show that participation and negotiation problems are connected to basic assumptions about markets. Expectations that credit developers will innovate and come up with new management practices to generate credits or that new areas will be designated as priorities mean that ESMs will keep changing. That makes it hard to stick to any agreement that is reached. As the efforts in the Chesapeake Bay states, the Midwest and the Pacific Northwest show, ESMs require ongoing negotiations about complex incentive structures.

Fung's work on public participation in a democratic context helps to illuminate many of the difficulties ESM designers face. He shows that "public participation at its best operates in synergy with representation and administration to yield more desirable practices and outcomes of collective decision making and action" (Fung 2006, 66). So, to reach a politically meaningful policy innovation like an ESM and implement it effectively, collective decision making and shared action are required. To enable such synergy, Fung suggests a "three-dimensional framework" based on key questions that can be asked of virtually any participatory process, including attempts to create of ESMs.

Who is invited?

An important theoretical feature of ESMs is that they create links between people and organizations that degrade or destroy ecosystems to landowners who have an option of managing their resources in a way that provides additional ecosystem services. If these two groups can find each other, they both stand to benefit. Landowners can gain financial benefits, while credit buyers can meet legal obligations at the lowest possible costs. How to help them find each other and how to involve them in the design of ESMs does not follow directly from these theoretical assumptions about market dynamics. It is typically impossible, for logistical reasons, to involve every landowner in order to convince them to consider developing ESM credits.

The distinction between regulatory and voluntary markets is one thing to consider in deciding whom to invite. If extensive laws and regulations govern a market, it is probably better to invite representatives of as many credit buyers and sellers as possible to participate in helping to draft those rules. If a market is voluntary, it suggests that fewer groups need to be involved in ESM design.

Most of the efforts in the Chesapeake Bay, the Ohio River basin and the Willamette River basin have been aimed at the creation of ESMs in the context of existing policies, like wetland mitigation banking[4] and Water Quality Trading.[5] These approaches tend to be characterized as *regulatory markets* (Freeman and Kolstad 2007). Within the broad category, the guiding hand of a government is assumed to operate in setting the environmental standards that need to be attained. Other expected participants in the creation of regulatory markets are landowners, since they have the potential to "supply" ecosystem services through the ecosystems present on their land. On the demand side of the market, regulated entities, like water utilities in need of an NPDES permit, are also expected participants, since they can be expected to potentially buy ecosystem services credits.

---

4 See for example these materials from the U.S. EPA: http://water.epa.gov/lawsregs/guidance/wetlands/workshops.cfm. Last accessed on 07/11/2015.

5 See this example: http://water.epa.gov/type/watersheds/trading.cfm. Last accessed on 07/11/2015.

Voluntary markets involve transactions between private entities unrelated to regulation. A well-known example of a voluntary market is the opportunity to purchase carbon credits when buying a plane ticket (Gössling et al., 2007). In the creation of voluntary markets, the involvement of government actors should not automatically be assumed, beyond perhaps guaranteeing that contracts are enforceable through the legal system. Other participants in voluntary markets are also less obvious, although landowners are still viewed as "service providers."

In creating the Nutrient Trading Program and Bay Bank in the Chesapeake Bay, the Ohio River Basin Trading Project and the Willamette Marketplace, the ESM designers were not always as explicit about the distinction between regulatory and voluntary markets. As described in Chapter 3 for both the Upland Prairie measurement system in Oregon and EPRI's nutrient credits, the market proponents tried to develop a measurement system for credit trading without the existence of a clear regulatory driver, meaning that no government agency was requiring an environmental impact to be compensated. The hope was that by creating a credit trading system in anticipation of a potential endangerment listing, for example, the regulators would be more likely to simply use that method to implement the Endangered Species Act. This makes identifying sources of demand, and thus knowing whom to invite, much less straightforward.

How do participants interact?

Inviting stakeholders to a meeting, or even a series of meetings, without a clear sense of what has to be achieved, who is authorized to make decisions and more generally how the group will work together, is unlikely to result in the outcome the proponents of ESMs desire. Process design in the field of environmental policymaking and planning is much discussed and debated (Carpenter and Kennedy 1988; Sabatier 2005; Susskind et al. 1999; Wondolleck and Yaffee 2000). After the question of who participates, political scientist Archon Fung considers a second dimension critical in his analysis of process design (Fung 2006). He has named this the *modes of communication and decision* (2006, 68). Different process designs can treat the public simply as listeners or spectators, or they can treat them as "experts" who can contribute perspective, expertise or experience. Finally, the public can be included as stakeholders who have a "right" to participate in negotiations.[6] It is not clear which is the best approach for the creation of ESMs.

---

6 The strengths and weaknesses of and the differences and similarities between various approaches to participation in decision making in the public sphere are at the center of a long-standing debate within professional and academic circles. An early contribution was Sherry Arnstein's 1969 article

Whether market proponents simply invite people to listen to an explanation of how an ESM is expected to operate or they invite participants to enter into detailed negotiations about the rules and regulations of a proposed ESM are two fundamentally different forms of interaction. The choice for one or the other, or some different form of interaction altogether, has practical implications, like how market proponents are likely to deal with reactions by or incorporate concerns of participants. Given the potentially significant outcomes of the creation of an ESM, like the concentration of investments in environmental restoration outside of a city's borders or the continued environmental degradation of specific places inside of them, the choice for a specific form of interaction also has more broadly democratic implications. How do residents who live near degraded streams or those who live near potential restoration sites engage with the creation of an ESM? Do they merely have to be informed that they can object to a permit that includes a trading provision, since it is uncertain if the development of a credit-generating restoration project will be close to them? Or should residents, community groups and advocacy organizations be actively involved in how and where to achieve environmental improvements? Do such groups even want to be involved in such discussions, or is it sufficient if the water treatment utility simply decides how to proceed on behalf of its ratepayers and informs them of its decision afterward? The uncertainties these questions pose, which I explain in more detail in this chapter, illustrate why ESMs are so difficult to create successfully, related to participation and negotiation. The Idaho Conservation League's open records request indicates that not all groups are going to agree with a very basic level of information provision.

The measurement systems in many ESMs are quite complex, and often highly technical, as described in the previous chapter. This makes it difficult for some participants to interact meaningfully in discussions about their design or use. So, even when a meeting or process is intended as an opportunity for a candid exchange of ideas, it is not always possible for all the participants to communicate effectively. In the creation of regulatory markets, a second source of profound complexity and miscommunication can emerge. This is the long history of specific laws and preexisting regulations that cannot be ignored. To use the example of wetland mitigation, this field originated with the Clean Water Act in 1972 and has seen a steady stream of regulatory decisions, legal

"A Ladder of Citizen Participation" in the *Journal of the American Institute of Planners*. More recently, the International Association for Public Participation (www.iap2.org) has developed a "spectrum" of participation methods, outlining different approaches to participation. This in turn has raised criticism, for example, for including information sharing as a form of participation, which can be considered a purely passive form of "engagement," rather than participation. See: Carson (2008).

actions and reactions, creating a long list of considerations and requirements for subsequent efforts to innovate in the wetland banking arena. Whether it is best to focus on specific details or search for agreement on broad principles in the development of an ESM related to wetlands is hard to determine. From an ideological standpoint, different stakeholders or participants will have competing views. From a technical standpoint, there are also likely to be diverse perspectives on this question depending on the disciplinary orientation of the individuals involved.

Finally, from a legal standpoint, even knowledgeable land use lawyers are likely to disagree on how much deference ought to be paid to existing rules and regulations. Efforts to simplify the agenda, in order to facilitate communication, can backfire. Concentrating on technical specifications only, and limiting who is invited to experts only, can result in implementation problems later on when certain stakeholder groups charge that they have been left out or that wrong assumptions have been made about their interests.

What kind(s) of authority do participants have?

The third dimension of participatory process design Archon Fung mentions is related to what he calls the *extent of authority and power* (Fung 2006, 69). Juxtaposing two extremes helps to illustrate the scope of this dimension. Does the outcome of an ESM design process reveal anything more than the preferences of the participants (who have no authority beyond perhaps changing their individual opinions)? Or, do key stakeholders have authority to implement what they design by virtue of their political or legal authority? The difference between a group engaged in brainstorming, which might result in little more than a report, and a rule-setting negotiation, in which outcomes are legally binding, is important when trying to create an institutional arrangement, let alone one as complex as an ESM.

Who has the power to set a market in motion? Should landowners simply begin restoration activities in the hope that buyers will find them, or should wastewater treatment plants contact individual landowners to persuade them to plant trees? Intermediaries, or so-called credit brokers, have often played an important role. In Pennsylvania, Red Barn Trading has served as a de facto representative of large groups of landowners, as have Soil and Water Conservation Districts in the Ohio River basin. Despite the existence of middlemen and informal representatives, moving ESMs from the drawing board to actual trading activity has turned out to be much more difficult than ESM advocates expected. The fact that no single organization or representative can simply force a market to form has been at the heart of this challenge.

4.b Nutrient Net and an Effort to Create Trading at the Bay Scale

The Chesapeake Bay watershed covers 64,000 square miles and (parts of) Delaware, Maryland, Pennsylvania, West Virginia, Virginia, New York and the District of Columbia. The sheer scale of the Chesapeake Bay watershed means that creating effective participatory processes to develop any kind of joint action covering the entire Chesapeake Bay watershed will automatically be complicated and almost inevitably somewhat disjointed. Despite this broad challenge, WRI received a Conservation Innovation Grant from the USDA in 2010 to develop a version of Nutrient Net, its online calculation and trading platform that is part of a measurement system for nutrient trading, covering the entire Chesapeake Bay watershed. The system was also supposed to include estimates for carbon dioxide offsets,[7] making this effort somewhat similar to EPRI's project in the Ohio River basin. This iteration of Nutrient Net was an attempt to move toward an ESM for the entire Chesapeake Bay watershed and was strongly supported by the U.S. EPA. It also encountered multiple negotiation and participation challenges.

Trading water quality credits across state boundaries in the Chesapeake Bay watershed is appealing since it creates the possibility for coastal states, like Maryland and Virginia, to extend their efforts to clean up the Chesapeake Bay beyond their boundaries. A significant improvement in water quality in the Chesapeake Bay would benefit Maryland and Virginia most directly, for example, through increases in commercial and recreational fisheries and coastal tourism (Talberth et al. 2010). Activities in states like Delaware and Pennsylvania impact water quality in the Chesapeake Bay, but since these states do not adjoin the Bay, there is less immediate concern about those impacts.

Such upstream/downstream dynamics, in which the jurisdictions at the mouth of a river suffer from the declining water quality caused by land uses and discharges upstream, are typical in many river basins and watersheds around the world. Water quality trades between upstream and downstream states would presumably allow considerable efficiency gains (Van Houtven et al. 2012). Pennsylvania's poultry farms contribute a great deal of nutrient pollution to the Chesapeake Bay. It is potentially much cheaper and more effective to achieve significant nutrient reductions related to poultry farming, than it would be to install even wider riparian buffer zones on agricultural land in Maryland, where a 100-foot buffer zone is already legally required. So when a water treatment utility in Maryland is allowed to buy credits from Pennsylvania, it can achieve more significant nutrient reductions at a lower

---

7 See: http://www.nrcs.usda.gov/wps/portal/nrcs/detailfull/national/programs/financial/cig/?&cid=stelprdb1044496. Last accessed on 08/06/2015.

Table 4.1 Water Quality Trading in Chesapeake Bay states

State	Registration Method	Credits	Status*
Maryland	Nutrient Net	Point Sources and Non-point sources	In development
Pennsylvania	Nutrient Net (until 2011) and Markit	Point source and Non-point sources	Active trading
Virginia	Individual Permits & Records held at VA Department of Environmental Quality	Point sources and Non-point sources	Active trading
West Virginia	Nutrient Net	Non-point sources	In development

* as of spring 2016.

Source: Branosky et al. 2011; www.deq.virginia.gov, www.mdnutrienttrading.com, www.dep.pa.gov, www.dep.wv.gov.

price than when its only credit options are in Maryland. At this point, since interstate trading does not happen, these price and effectiveness differences are largely speculative, but extensive research has shown that nutrient reductions in Pennsylvania are potentially very effective, at low cost (Van Houtven et al. 2012).

WRI, together with the Chesapeake Bay Program, has worked to expand Nutrient Net and to facilitate buying and selling of water quality credits across state boundaries by integrating the various state trading programs that exist or are under development (Table 4.1).

So far, these efforts have not actually resulted in an integrated ESM for the Chesapeake Bay. The profound challenges associated with developing an effective participatory process to create an ESM at this scale may not be possible to overcome.

Participation and communication to what end?

One of the basic problems of organizing effective participatory processes in the Chesapeake Bay watershed is that there is already a myriad of committees, working groups and teams involving somewhat overlapping, yet surprisingly disconnected efforts and not enough clear pressure to make progress on key environmental goals. There are so many committees that many experts and representatives find it difficult to be present at the many meetings, workshops and procedures involved. In addition, coordination of efforts across these groups represents a substantial management problem. In Archon Fung's terms, participant selection and effective (modes of) communication are even more complicated as the number of autonomous jurisdictions involved increases.

Two prominent experts on environmental issues in the Chesapeake Bay have described the long-standing oversupply of committees, working groups and teams as *Collaborating to Nowhere*, framing this as a government accountability problem (Steinzor and Jones 2013). In their argument, it is not in the immediate interest of the upstream states, especially Pennsylvania, to reduce nutrient run-off dramatically. So until the federal government, through the EPA or by appointing an independent evaluator (which Steinzor and Jones much prefer and describe in some detail), creates a clear incentive for Pennsylvania to meaningfully participate, or a penalty for nonparticipation, collaboration and negotiation are unlikely to achieve meaningful results. While it falls outside of the scope of this book to evaluate this claim, it is relevant to point out that the history of attempts to collaborate and negotiate across state boundaries in the Chesapeake Bay is long and controversial, and the attempts to create an interstate ecosystem service market can be seen both as a response to this problem (by seeking to create financial incentives for nutrient reductions in Pennsylvania) as well as adding to it (by creating yet another series of workshops and committees without centralized oversight or strong authority).

The two organizations that are central in the efforts to develop Water Quality Trading in the Chesapeake Bay states at the scale of the entire watershed are the Chesapeake Bay Program and the World Resources Institute. The Chesapeake Bay Program (CBP) is the most important watershed partnership in the region.[8] It is led by the governors of Maryland, Pennsylvania and Virginia, the mayor of the District of Columbia, the EPA administrator and the chair of the Chesapeake Bay Commission through the Chesapeake Executive Council. The CBP grew out of a 1983 agreement among these organizations and its 83 staff members are housed in an office in Annapolis, Maryland. The not-for-profit World Resources Institute (WRI) has long played an important role in the promotion of Water Quality Trading in the Chesapeake Bay states, in part because of its work on the development of Nutrient Net.[9] WRI has provided important logistical and technical support to Water Quality Trading programs in Pennsylvania, Maryland and West Virginia, although the latter two have not yet seen any trading activity. WRI seeks to enable interstate trading based on its belief that matching supply and demand requires significant scale (Selman et al. 2010).

---

8 For a recent, highly critical analysis of the history of the Chesapeake Bay Partnership, see Steinzor and Jones, "Collaborating to Nowhere: The Imperative of Government Accountabiliy for Restoring the Chesapeake Bay." *Journal of Energy & Environmental Law*, Winter 2013.

9 For Pennsylvania's portal, see: pa.nutrientnet.org. Last accessed on 08/30/2015.

The participants in WRI's two-year effort to develop a Bay-wide Nutrient Net were primarily state agencies that run the Water Quality Trading programs in the Chesapeake Bay states, except West Virginia and Delaware where there was no interest in trading at the time. These two states chose not to participate for different reasons. West Virginia already had a formal trading policy that was not being used. The state was less than enthusiastic about devoting additional resources to expanding it. Delaware on the other hand has typically adopted Maryland's policies and tools, so the expectation is that whatever ESM design Maryland ends up implementing, Delaware will simply follow. Most of the land in Delaware does not drain into the Chesapeake Bay, making an ESM focused on that water body a minor concern for the state. Other participants in the Bay-wide Nutrient Net process were two private mitigation bankers from the region, the USDA's Office of Environmental Markets and the Chesapeake Bay Program's policy specialist on trading and offset programs (WRI staff member, interview with author, November 15, 2012). The group of stakeholders seems quite small given the number of organizations and people with a direct stake in water quality when compared to the efforts in Oregon and the Ohio River basin.

Highlighting the complexity of the collaborative efforts related to Water Quality Trading among states is the fact that shortly before WRI began its effort to develop the Bay-wide version of Nutrient Net, President Obama issued Executive Order 13508, titled *Strategy for Protecting and Restoring the Chesapeake Bay Watershed*. As part of this order, an interdepartmental *Chesapeake Bay Environmental Markets Team* was created.[10] This team consisted of federal agencies involved in the conservation and restoration of the Chesapeake Bay led by USDA's Office of Environmental Markets. WRI did not participate directly. It is not clear if and how these efforts were coordinated or mutually supportive.

The most significant development related to trading water quality credits in the Chesapeake Bay watershed, coinciding with WRI's effort, was the release of the Chesapeake Bay Total Maximum Daily Loads (TMDL) by the EPA in December 2010. The agency describes it as follows:

The TMDL—the largest ever developed by EPA, encompassing a 64,000-square-mile watershed—identifies the necessary pollution reductions from major sources of nitrogen, phosphorus and sediment across the District of

---

10 The *Strategy for Protecting and Restoring the Chesapeake Bay Watershed* was released in 2010, and was a response to Executive Order 13508. For information about the order, the order itself and a report on the progress made, go to: http://executiveorder.chesapeakebay.net/page/About-the-Executive-Order.aspx. The Strategy is available at: http://executiveorder.chesapeakebay.net/file. axd?file=2010%2f5%2fChesapeake+EO+Strategy.pdf and the Charter of the Environmental Markets Team can be found at: http://www.usda.gov/oce/environmental_markets/files/FY10_ 11_CB_EMT_Charter_final.pdf. All retrieved on 08/08/2012.

Columbia and large sections of Delaware, Maryland, New York, Pennsylvania, Virginia and West Virginia, and sets pollution limits necessary to meet water quality standards in the Bay and its tidal rivers.[11]

This means that the EPA now had the authority to review the Water Quality Trading programs in the Chesapeake Bay watershed. States are required to show how their trading programs are likely to meet the standards set by the EPA's TMDL. The Chesapeake Bay Program, led by EPA, developed a forum to discuss trading related to Water Quality Trading following the release of the TMDL. It was called the Trading and Off-Sets Work Group (TOWG).[12] Formalized in 2011, it laid out an ambitious work plan.[13] WRI is a member of this group, which includes more than 50 representatives of environmental advocacy organizations, state and federal agencies and scholars.[14]

For the participants in WRI's process to create a single Nutrient Net, these simultaneous and somewhat overlapping processes meant no single forum existed to discuss issues related to technology development or policy coordination for Water Quality Trading across all Chesapeake Bay states. This caused delays and made it hard to schedule meetings. The effort to create a single version of Nutrient Net for the entire Chesapeake Bay watershed was not finished when the USDA grant ended. Additional funding had to be sought.

Questionable authority in the Chesapeake Bay

The development of Nutrient Net was oriented toward building appropriate measurement systems and creating a shared understanding of state trading schemes, rather than explicit consensus on policy questions. The fragmentation of discussions and decision making regarding market creation in the Chesapeake Bay watershed, despite decades of experimentation with Bay-wide institutions like the Executive Council, and now the Trading and Off-Sets Work Group and the Chesapeake Bay Environmental Markets Team, means that deliberations and authority remain dispersed.[15]

---

11 For an overview of the TMDL and associated activities, see: http://www.epa.gov/chesapeakebay-tmdl/. This quote is from the factsheet, available at: http://www.epa.gov/reg3wapd/pdf/pdf_chesbay/BayTMDLFactSheet8_6.pdf. Both retrieved on 08/09/2012.

12 For this group's history and charter, see: http://www.chesapeakebay.net/groups/group/trading_and_offsets_workgroup. Last accessed on 08/23/2015.

13 See: http://www.chesapeakebay.net/channel_files/18622/draft_towg_work_plan_08-3-12.pdf. Last accessed on 08/08/2015.

14 For the list of participants, see: http://www.chesapeakebay.net/groups/group/trading_and_offsets_workgroup. Last accessed on 08/23/2015.

15 The best place to find a description of these specific committees is the Chesapeake Bay Program's website: http://www.chesapeakebay.net/about/organized. Last accessed on 08/13/2015. A more

The extent to which the EPA and the Trading and Off-Sets Work Group will be able to force changes to individual state trading systems is unclear. As an official from the agency put it: "It remains to be seen (…) how far reaching the charge and impact of this group would be" (Interview with author, November 15, 2011). He went on to say more about limitations on the group's authority:

> So I can't say from that angle what kind of things this group would take on and frankly how much—I use a political science term here, but how much sovereignty any of the states would be willing to give up in that process and how much direction they'd be willing to take. You know, by extension one has to ask how much political will does EPA or any other federal agency have to try to compel the states to comply with the determinations of a group like this. (Interview with author, November 15, 2011)

For now, interstate trading of water quality credits remains practically impossible. It is unclear how or why this might change as a result of Nutrient Net. The examples in the Ohio River basin and Oregon, where much energy and resources were spent developing market infrastructure and measurement systems without clear regulatory drivers, do not bode well for the market proponents in the Chesapeake Bay.

A second complicating factor reveals still another challenge, more specific to the development of ESMs and the effort by WRI to create a single Nutrient Net for all the Chesapeake Bay states. A senior associate at the World Resources Institute and the project lead on multiple iterations of Nutrient Net describes the relationship between technology development and policy advocacy as follows:

> WRI isn't an advocacy organization, but we are an environmental think tank. (…) It's delicate, because by building the tool for the states it is almost—we are never anybody's—we are not the contractor, they are not our clients, et cetera, although sometimes it becomes like that. (Interview with author, November 15, 2011)

This ambivalent description of the relationship between WRI and the responsible state agencies points to the difficulties surrounding ESM development and participation. It reveals uncertainty with regard to the roles and responsibilities of the institutions involved. Many groups want to participate in the

critical analysis of the history of collaboration and institutional design in the Chesapeake Bay is: Steinzor and Jones, "Collaborating to Nowhere."

development of markets for ecosystem services, but their involvement in specific working groups, focusing on technology development or rule making, for example, does not constitute a commitment to the implementation of markets in which these technologies and rules might be used.

There is an irony in the fact that the emergence of numerous working groups and teams has made communication more difficult, thereby paradoxically limiting meaningful participation opportunities. Some organizations and individuals have simply not been able to attend required meetings. Lines of authority are confusing or nonexistent. Neither the EPA nor any other organization has the authority to implement a Bay-wide ESM, focused on water quality or any other type of environmental credit. The long history of efforts to clean up the Chesapeake Bay suggests this should come as no surprise. The much more limited authority of an organization like WRI, which has long advocated Water Quality Trading for the entire watershed and is an important actor with great standing in the field, is not clear either. While active participation by organizations like WRI brings with it many resources and expertise, it also raises the question of where the authority ought to be located when ESMs cross state and agency boundaries.

4.c EPRI and the Role of Soil and Water Conservation Districts

In a newsletter issued during the summer of 2010, EPRI released an elaborate organizational diagram to structure the participation of a range of stakeholders in the development of the proposed ESM in the Ohio River basin.[16] This organizational structure included subcommittees on the power industry, municipal wastewater treatment plants and agriculture and environmental groups. The level of detail and forethought this diagram represents is indicative of EPRI's approach to participation in the process of creating an ESM in the Ohio River basin. In sharp contrast with the overlapping and often confusing task forces, working groups and processes that characterize efforts in the Chesapeake Bay watershed, EPRI developed a clearly structured process that the organization hoped would result in a widely supported ESM.

Despite this sophisticated participatory decision-making process, important stakeholders reduced their level of participation over time and some expressed concerns over the program through channels external to EPRI's committee structure. To fully appreciate both EPRI's significant investment in participation and the failure to keep all relevant stakeholders engaged and supportive

---

16 This newsletter was retrieved from http://wqt.epri.com/schedule-updates.html on 07/18/2015.

throughout the process, a closer look at two of the three characteristics Fung outlined is necessary, namely *mode of communication* and *decision-making authority*.

The first section focuses on the way in which participants communicated and made decisions within the context of the participatory process in the Ohio River basin project. In the development of ESMs more generally, an important question is whether organizations can influence not only the design of an ESM but also its implementation on the level of individual credit-generating projects and transactions. A number of environmental groups raised this concern, and it represents a fundamental challenge to the ESM concept. While basic information about individual projects is typically made available, the basic idea of the market is that transactions are made between buyers and sellers, without much second-guessing by interested outsiders like environmental groups or local stakeholders. EPRI's experience in the Ohio River basin shows that this question cannot be readily answered to the satisfaction of all relevant stakeholders.

The second characteristic that is of relevance to participation in the creation of ESMs, the *decision-making authority*, relates to the question of how a process led by a nongovernmental organization like EPRI can influence decisions outside of the normal regulatory process. Many ESM proponents operate outside of the formal regulatory structure, meaning they work in nongovernmental organizations but frequently rely on monetary support from the USDA and the EPA, through various grant programs. Officials from these and other agencies typically participate actively in the processes designed to create ESMs. Does this mean that the outcomes become, or significantly inform the regulatory standard? This ambiguity is characteristic of many complex environmental governance processes (Innes and Booher 2010).

The idea that a nongovernmental actor can independently develop a framework that is widely seen as a legitimate form of environmental policy-making has been described as the *emergence of non-state authority* (Cashore, Auld and Newsom 2004). This idea emerged in the context of another type of market-based ecosystem services initiative, namely the creation of certification systems that promote environmentally sustainable practices in the timber industry in Sweden. These authors describe this phenomenon with the intriguing phrase *governing through markets*. How the possibility of *governing through markets* can become problematic from the perspective of those who don't agree with the creation of an ESM, I will describe next based on EPRI's work in the Ohio River basin.

Dealing with resistance

In the spring and summer of 2012, EPRI brought multiple national and Midwestern environmental advocacy organizations together during biweekly conference calls to discuss the development of an ESM in the Ohio River basin. Following this intensive interaction phase, more irregular communication between EPRI and key members has continued via phone calls and more generally via newsletters and webcasts. Mirroring the conversations between these environmental organizations were similar ones between a group of organizations representing agricultural interests and a group made up of electric power companies. The final category of potential stakeholders in this ESM was the wastewater treatment sector, which was represented through two umbrella organizations, the Ohio River Valley Water Sanitation Commission (ORSANCO) and the National Association of Clean Water Agencies (NACWA).

The type of interactions and mode of decision making among the participants in this process, be it majority-based voting, seeking full consensus or something in between, was not immediately clear. In its original organization diagram, EPRI referred to the sectoral groups as *subcommittees* to the *stakeholder steering committee*, which in turn reported to the *project team collaborators*. After some initial interactions, all *subcommittees* were renamed *advisory committees*, indicating a subtle adjustment of the way in which participants talked about the type of influence of these groups. A member of the environmental advisory committee described this shift as justified. His reasoning was that the word "advisory" more accurately described the level of influence on the outcome of the project than the word "steering" (Interview with author, April 2, 2015).

The members of the *environmental advisory committee* were both national and state environmental groups, namely the Ohio Environmental Council, the Nature Conservancy, the Sierra Club, the Environmental Law and Policy Center, the Minnesota Center for Environmental Advocacy, the Kentucky Waterways Alliance and Environmental Defense. Many of these groups collaborate on water quality issues in the Mississippi River Collaborative[17] created in 2005. Under this umbrella, these environmental groups have been trying to force the EPA to set nutrient standards for the entire Mississippi drainage area, including the Ohio River basin, through legal action.

The oldest environmental organization among these participants, the Sierra Club, has long been highly critical of Water Quality Trading. As early as 2002, in response to the EPA's draft rule on Water Quality Trading, the Sierra Club expressed concerns over displacement and nonequivalence:

Pollution credit trading can create pollution "hot spots" where dischargers can dump more pollution simply because they have bought credits. Under the proposal, dischargers can trade credits between different areas and

17 See: http://www.msrivercollab.org. Last accessed on 09/01/2015.

among different pollutants. As a result, pollution problems merely shift from one area to another. (Sierra Club 2002, 4)

Given this track record of resistance, the Sierra Club's participation in EPRI's advisory committee is both remarkable and highly relevant to this analysis of participation in the design of an ESM.

After following the project's development closely and participating in meetings for close to two years, the Sierra Club's representatives in the Ohio River basin remained deeply concerned about EPRI's trading scheme. In 2014, Sierra Club chapters in Ohio, Indiana, Illinois and Tennessee expressed the problems they saw in a direct letter to the responsible senior EPA official. This letter raises a number of issues, including strongly worded doubts that the WARMF modeling an EPA Region 5 spreadsheet estimates can produce conclusive evidence regarding the equivalence between polluted discharge and BMP installation on upstream farms. More directly relevant to the challenge of participation in the creation of an ESM, however, are two issues raised in the letter, namely the role of Soil and Water Conservation Districts in Best Management Practice installation and the influence of EPRI's pilot program on potential regulatory decisions.

Soil and Water Conservation Districts (SWCD) play a central role in EPRI's ESM, since they identify opportunities for, oversee installation of and eventually verify the effectiveness of all credit-generating Best Management Practices. These districts typically operate at a county level and emerged in response to the Dust Bowl of the 1930s.[18] These organizations are established under state law and generally provide assistance to farmers by helping them protect soil productivity and water quality. In many counties with a large agricultural sector, Soil and Water Conservation Districts are an important source of engineering, financial and regulatory expertise that farmers draw on. The areas in Ohio, Kentucky and Indiana EPRI identified as potential locations for its first rounds of credit generation are no exception.

In the Ohio River Basin Trading Program, officials from the Soil and Water Conservation Districts identify potential farms where credit-generating BMPs could be located. After EPRI reviewed these proposed sites and estimated the nutrient reductions that might be achieved as a result, the SWCD would then oversee the installation of the fence, cover crop or heavy use pad for cattle to prevent erosion and thereby reduce nutrient run-off. Once this equipment has been installed, those same SWCD officials are responsible for the ongoing on-the-ground verification of the continued maintenance of those BMPs. This

---

18 For a brief history of conservation districts, see the National Association of Conservation Districts website: http://www.nacdnet.org/about/districts. Last accessed on 09/02/2015.

is obviously a labor-intensive process and the Soil and Water Conservation Districts are reimbursed for this work in EPRI's ESM, since they retain 10% of the cost of credit development as overhead. Since EPRI set a maximum cost for each project at $10,000 for the first two rounds of credits, SCWD received about $1,000 per project.

The Sierra Club's letter to the EPA raised multiple, related concerns over this model. It is therefore worth quoting at some length:

> Finally, all trading programs must be transparent to the public and comply with all of the Clean Water Act's public participation requirements. The public should be able to identify exactly where these practices are being installed and how much nutrient reduction they are estimated to achieve. The use of soil and water conservation district officers rather than employees of regulatory agencies to confirm that management practices have been installed and are operating correctly is worrisome. Farmers who wish to profit from nutrient criteria imposed upon regulated dischargers should be subject to the same regulatory scrutiny that those dischargers are.[19]

This type of concern, over who gets to inspect the restoration practices and what kind(s) of public input are required or allowed about individual credit-generating projects are not unique to the Ohio River basin's trading scheme, nor are they exclusively expressed by the Sierra Club. The fact that the organization felt that the *advisory committee* was not a sufficiently influential venue to express these issues reflects an important element in the challenge of developing a participation process for ESMs. While centralizing participation and negotiation is possible during the design of an ESM, once the market is operational, numerous credit projects and transactions will take place simultaneously, at least in theory. The potential volume of those projects and transactions makes effective public scrutiny difficult, even if much information is technically accessible through a web platform like Markit's Credit Registry, where EPRI posts relevant data on its credit-generating projects. In the concluding section to this chapter I will return to this difference between transparency and scrutiny.

Committing state regulators

An important goal of EPRI's Ohio River Basin Trading Project was to develop a market that would allow credit buyers to purchase water quality

---

19 This paragraph appears on page 3 of the letter from the Sierra Club to Nancy Stoner, sent on March 14, 2014. I received this letter through personal correspondence with Bowden Quinn, conservation director of the Hoosier Chapter of the Sierra Club, and one of the people who signed the letter.

credits across state lines. Ohio was the only state with existing guidance on how to develop a Water Quality Trading scheme among those where EPRI was interested in developing nutrient credits in its pilot phase. This made the Ohio EPA an important early participant in EPRI's development process. The agency supported EPRI's attempt to expand Water Quality Trading in the region and its willingness and ability to invest in the development of more sophisticated measurement systems. With Ohio's basic trading framework codified in the Ohio Administrative Code (Chapter 3745-5) and Ohio EPA's experience with trading in the Great Miami River and Sugar Creek watersheds, this state was unlikely to fundamentally alter its long-standing approach based on this new interest. Given EPRI's interest in facilitating trades across state boundaries, it was necessary to convince relevant officials in Indiana and Kentucky, the other two states where EPRI focused much of its attention, to accept trading rules and guidelines in line with those in place in Ohio.

On August 9, 2012, senior agency officials from all three states signed a trading plan that enabled interstate trading during EPRI's pilot project, showing EPRI's influence. This announcement was reported locally in Indiana, Ohio and Kentucky, and nationally in some water- and agriculture-specific outlets. The responses were not uniformly positive, and concerns generally focused on the issues of displacement and nonequivalence. The *Courier-Journal* in Louisville reported that a local environmental group, the Floyds Fork Environmental Association, was concerned about this program simply shifting pollution from one spot to the other. The article also mentioned environmental justice concerns, as low-income and minority populations frequently live close to heavy industries that might continue to pollute their immediate environment under a trading scheme. This article also cited one of the environmental groups' representatives on the advisory committee, the Kentucky Waterways Alliance, arguing that the main reason this organization was involved was to influence the specific outcomes in Kentucky, since the arrival of Water Quality Trading was inevitable.

It was exactly this question about the ability to drive agencies' decision making prior to regulatory implementation that EPRI itself had mentioned as an important reason for electricity companies to support the pilot project. In its business case for the electricity industry, EPRI had stated:

> Working with other key stakeholders that appear ready for a regional trading program will also reduce political or institutional risks. This will yield dividends to the power industry in terms of goodwill, public relations, and active participation in development of nutrient criteria implementation plans by ORSANCO and state agencies. (EPRI 2010, 5–2)

This kind of influence, or at least the perception that EPRI was able to drive states to accept specific approaches to Water Quality Trading, made several environmental organizations uncomfortable. The Sierra Club, in its 2014 letter to the acting assistant administrator at the U.S. EPA made its discomfort very clear:

> In the absence of an effective regulatory "cap" to trade under, the EPRI program has offered various inducements for dischargers to buy credits. These inducements include promises of "permit flexibility" and the use of stewardship credits as supplemental environmental projects to initial credit buyers. We believe it is important for all states and regulated parties to understand that any changes to a state's NPDES permitting program (including the offering of "flexibility" and credits towards NPDES permit compliance) must be validly approved through a state's public rulemaking process and approved by the U.S. E.P.A. Emerging trading programs should be careful about making promises that they can't deliver as an inducement to participate in the program. (Personal communication with author, April 2, 2015)

From the perspective of the state regulators, this kind of concern seemed unfounded. EPRI's project clearly occurred outside of the scope of formal regulatory action, and the sector EPRI was most closely associated with, the electricity industry, did not face any immediate regulatory drivers for nitrogen in this region.

The question of whether EPRI has been able to meaningfully influence state and federal regulators through its Ohio River Basin Trading Project remains difficult to answer. On one hand, the main regulatory action of direct relevance to EPRI's members is the EPA's Steam Electric Effluent Limits Guideline, which is not in line with EPRI's goals for Water Quality Trading. On the other hand, numerous state and federal officials have expressed, formally and informally, their support for EPRI's work and general approach to credit generation, calculation and verification. While it's impossible to say if and how elements of EPRI's work will be incorporated into Water Quality Trading in the Ohio River basin, it's clear that EPRI's attempt to create an integrated ESM has not yet succeeded in the way envisioned just a few years ago.

In summary, while EPRI designed and implemented a clearly structured and well-organized participatory process to create an ESM in the Ohio River basin, significant concerns remain about the roles and relationships of the different organizations that have an interest in ESMs. One is about the ongoing scrutiny some environmental groups believe is necessary and the other is about the appropriate relationship between implementation of existing regulations and innovative approaches like Water Quality Trading. The fundamental nature of these concerns and the inability to resolve them conclusively during EPRI's effort to create an ESM raises the question if any level or type of participatory process can achieve a broad agreement about an ESM. To answer this question, we now turn to Oregon, where consensus-based approaches have long been an important element of environmental policymaking, and the Willamette Partnership pursued its answer most ambitiously.

4.d Seeking Consensus in Oregon

Determining who should participate in the design and implementation of an ESM, how stakeholders should communicate and what influence decisions made by the group have on regulatory outcomes are significant barriers to the realization of markets for ecosystem services. If there were one effort to create an ESM that seemed likely to answer these questions in a successful way, it would be the Willamette Partnership's attempt to create the Willamette Marketplace in Oregon.

The Willamette Partnership and Clean Water Services, the water utility that received the first shade credit–based NPDES permit, together organized a yearlong consensus-building process from 2008 until 2009 with the goal of creating an ESM. They called this effort Counting on the Environment (COTE).

The following sections provide a more in-depth look at the specific problems related to participation that emerged during the COTE process. The goal of this analysis is to highlight the profound challenges of bringing stakeholders together to design and implement ESMs, even in a context where the promise of success seems overwhelming. The Willamette Partnership and Clean Water Services had secured significant funding for this process through a Conservation Innovation Grant from USDA[20] and used this money in part to bring in an expert in building consensus on complex environmental policy issues.[21] Regulatory agencies were well represented and seemed committed to the process and its outcomes. One of the principal actors in the process, the lead representative from Clean Water Services, has written a PhD dissertation on the importance of networks in the creation of Water Quality

---

20 For a brief description of the original goals and size of the grant, see the NRCS website: http://www.nrcs.usda.gov/wps/portal/nrcs/detailfull/?ss=16&navtype=SubNavigation&cid=stelprdb1044500&navid=100120290000000&pnavid=100120000000000&position=Not%20Yet%20Determined.Html&ttype=detailfull&pname=2007%20Conservation%20Innovation%20Grants%20Awards%20|%20NRCS. Last accessed on 05/19/2015.

21 See this description of the Counting on the Environment process for more information on the role of the mediator: http://www.oregonconsensus.org. Last accessed on 05/19/2015.

Trading schemes, drawing in part on material from the Great Miami River basin in Ohio (Cochran 2008). In short, all the ingredients for a successfully executed ESM design and implementation process seemed present in Oregon. Yet beyond the concerns over displacement voiced by the City of Portland and the National Marine Fisheries Service described in Chapter 2, two key concerns over participation emerged.

The first concern was over the basic question: *Who is invited?* The initiators of the process from the Willamette Partnership and Clean Water Services describe their participant selection strategy for COTE in terms of "building support" for ESMs. That's a very specific orientation toward stakeholder selection and engagement. It raises the question of whether they were open to as wide an array of opinions on markets and market creation as they could have been. This approach would later result in severe criticism from a number of prominent environmental advocacy organizations in the state.

The second challenge related to participation was the role of expertise, a returning issue in efforts at collaborative decision making in complex environmental disputes (Ozawa 1991). The question at the center of this problem is how expertise can be brought to bear on decision making when participants have a diverse range of backgrounds in terms of disciplinary training, professional knowledge and skills. In the Counting on the Environment process in Oregon, many different types and levels of expertise were represented, as is to be expected in the development of an ESM. Despite a deliberate process design and much emphasis on the collaborative development of measurement tools and instruments, the disciplinary differences remained an obstacle. The following sections will describe these two issues in more detail, before the conclusion of this chapter returns to the broader question of why the structure and extent of participation in the design and implementation of ESMs is the third obstacle to their successful realization.

Counting on support

A widely accepted but labor-intensive way to start the development of a participatory decision-making process is by conducting a stakeholder assessment or analysis (Reed et al. 2009; Rockloff and Lockie 2004; Schenk 2007; Susskind, McKearnan and Thomas-Larmer 1999). This type of analysis consists of developing a list of possible stakeholders and describing their key interests and concerns on the topics of discussion. Part of the goal of this exercise is to ensure that all relevant stakeholders are engaged in the participatory decision-making process, or at the very least made aware of it. When an external, more independent expert is brought in to facilitate a complex consensus-building process, he or she typically manages this information-gathering exercise

(O'Leary and Bingham 2003). In the initial phase of what would become the COTE process, this analysis was conducted in a different way.

The Willamette Partnership and Clean Water Services suggested relevant stakeholders, and the environmental mediator then conducted phone interviews with specific individuals from those organizations to gather more detailed information. This initial list of stakeholders only included supporters of the idea of an ESM. That was the result of a clear decision by Clean Water Services and the Willamette Partnership:

> We wanted people that we knew didn't have a fundamental knee-jerk reaction against mitigation or market-based approaches. We didn't want people to all be supportive, but we didn't want people just philosophically opposed. (Interview with author, July 8, 2011)

Starting with a list of suggested stakeholders provided by the organizations funding a consensus-building process is not uncommon, but frequently the list of relevant stakeholders expands during the assessment. That did not happen in this case, which is surprising given the track record of some of the environmental advocacy groups in and around Portland.

The Northwest Environmental Defense Center (NEDC), for example, did not participate in the COTE process, nor was it invited to. This organization has been in existence since 1969 and is closely connected to the Lewis and Clark Law School in Portland, Oregon. Its achievements include a significant legal victory against Clean Water Services in 1990, when that organization was still called the Unified Sewerage Agency of Washington County. NEDC filed the lawsuit based on its claim of repeated violations of NPDES permits at several wastewater treatment plants in the Tualatin River basin. What made this lawsuit particularly meaningful was the fact that its settlement[22] required the creation of a "Tualatin River Basin Water Quality Endowment" into which CWS would contribute at least $900,000 and possibly more in the case of additional violations. At the time, it was one of the largest settlements a municipal agency had ever been forced to pay.[23]

The organizers' desire to invite only representatives of organizations without an existing dislike of market-based environmental policy instruments might seem logical, given the explicit goal of the COTE process to reach consensus on an implementable ESM. On the other hand, the NEDC remains

---

22 The case is known as *NEDC v. Unified Sewerage Agency of Washington County*. The full text of the consent decree was retrieved from: http://www.trwc.org/water/docs/dfiles.pdf on 08/15/2015.

23 See: https://law.lclark.edu/centers/northwest_environmental_defense_center/projects/water.php Last accessed on 08/15/2015.

an active and vocal member of the advocacy community around water quality issues in Portland, where most of the other participants in COTE are located as well. An organization that frequently works with the NEDC, the Northwest Environmental Advocates, has also been excluded from the Willamette Partnership's efforts to create an ESM. As that organization's executive director describes:

> The Willamette Partnership, in my view, has never reached out to the organizations that have been known for doing Clean Water Act advocacy work over the years. Not only Northwest Environmental Advocates. (Interview with author, April 30, 2015)

When asked about why, she goes on to describe how she believes the Willamette Partnership and especially its frequent partner organization, the Freshwater Trust, regard her organization and others like the NEDC:

> This was not just because of an omission, overlooking the role of the law and the organizations that work with the law, but a rather great disdain for them. (Interview with author, April 30, 2015)

NWEA, beyond filing the open records request under the Freedom of Information Act (FOIA) mentioned in the opening paragraph of this chapter, together with the Idaho Conservation League and NEDC, has started to follow the development of the Willamette Partnership's efforts to create ESMs very critically. For example, in 2011 the City of Medford received the second NPDES permit in Oregon in which shade trading fulfilled part of its obligation to cool treated wastewater.[24] NWEA wrote a highly detailed and very critical letter to the EPA[25] in response to that permit, challenging the overall legality of Oregon's approach to Water Quality Trading. The letter argued Oregon's Department of Environmental Quality (ORDEQ), which issued the permit, should revise its permitting approach completely when it comes to Water Quality Trading. Shortly after NWEA sent this letter, the Department announced that it would develop new rules for Water Quality Trading, a process that will take more than a year to complete. In the interim, no new trading based permits will be issued in Oregon.

It's impossible to prove whether inviting NWEA and/or NEDC to the Counting on the Environment process could have reduced their opposition

---

24 See: http://www.deq.state.or.us/wq/trading/trading.htm. Last accessed on 09/07/2015.
25 The letter can be accessed here: http://www.northwestenvironmentaladvocates.org/nweafiles/2013/03/Letter-to-EPA-on-Oregon-trading-3-2013.pdf. Retrieved on 09/08/2015.

to ESMs or altered the approach to Water Quality Trading implemented in Medford sufficiently to avoid NWEA's criticism. However, the letter to the EPA about the Medford permit does repeatedly invoke the original shade-trading permit, issued to Clean Water Services back in 2005, as a good example of Water Quality Trading. This casts doubt on the notion that NWEA is fundamentally opposed to trading, making it possible that its inclusion in COTE could have resulted in a productive dialogue.

The decision to invite stakeholders without a prior (perceived) antipathy toward market-based approaches to environmental policy issues and the ensuing opposition from groups like NWEA and NEDC highlight the difficulty of designing and implementing an effective participatory decision-making approach to create an ESM. In the Ohio River basin, EPRI actively engaged the Sierra Club, an organization known as highly critical of Water Quality Trading. This did not prevent its representatives from forcefully voicing their concerns and complaints directly to the U.S. EPA. In Oregon, the Willamette Partnership and Clean Water Services decided to avoid including organizations with a more critical stance, despite taking advantage of existing expertise in process design in the form of a neutral mediator with experience handling acrimonious environmental policy debates. This exclusion of certain environmental groups did nothing to prevent, and possibly exacerbated, forceful criticism of the trading scheme further developed in COTE. NWEA's severe criticism of that trading scheme might have contributed to the responsible agency temporarily halting the issuance of all trading-based permits.

Given the limited number of efforts to create ESMs of the scale and scope of the Willamette Marketplace and the Ohio River Basin Trading Project, it is hard to say if it's impossible to more productively engage environmental advocacy organizations with deep concerns over trading per se. At this point however, deciding if and how to engage critical voices is a profound challenge to the successful creation of ESMs in the United States. But perhaps more distressing to their proponents, COTE also revealed that it is hard to keep those supportive of market-based approaches productively engaged during the process of creating markets for ecosystem services.

Communicating across boundaries

During the COTE process, as is typical for the creation of an ESM, the participants spent a lot of their time discussing measurement systems and the production of equivalence. These conversations and negotiations touch on one of the basic challenges of the ESM concept but also present an obstacle specific to meaningful participation. This challenge is how to have a productive conversation about sophisticated measurement instruments when the participants have different disciplinary and professional backgrounds, with varying levels of training or familiarity with computer models.

This question is clearly not unique to ESMs. In fact, the challenge of "common representation in diverse social worlds" (Star and Griesemer 1989) and the search for objects that can span the boundaries between those social worlds is relevant far beyond environmental policy issues. Such a *boundary object* can be as simple as a standardized form or an artifact like an atlas, as long it enables the spanning of the various boundaries between types and levels of expertise. However, Star and Griesemer's now classic analysis shows how difficult it is to create ways of interacting between amateurs, professionals and experts from various disciplines. While the measurement systems central to the efforts to produce equivalence for ESMs can be seen as attempts to create boundary objects, the experiences in the Ohio River basin and Oregon in the previous chapter indicate how difficult that is.

In the COTE process, this resulted in increased disengagement from the overall exercise by certain participants. These individuals did not oppose the creation of ESMs in general, or even fundamentally disagree with the claims of equivalence made based on the specific measurement systems. Instead, these representatives believed the proponents' inability to cross the boundaries of the ecological origins of the ESM approach prevented them from meaningfully participating. One participant from Oregon's Department of Land Conservation and Development described this dynamic as follows:

> I will say that from the beginning, it was very scientifically oriented, and so I did a lot of listening at the beginning, and I felt a little bit apart from the others there who had more of a science background, and my background is more land use—just land use planning. (Interview with author, July 18, 2011)

And while this representative did raise a number of issues specific to her disciplinary background, in this case land use planning, she did not believe these were taken seriously enough.

In the end, she believed the main insights from her point of view were largely absent from the ESM overall: "So it seemed to me that planning and zoning, proximity to UGBs [Urban Growth Boundaries, TVM] and city limits should have been in, should be in there." This absence resulted in her conclusion that the ESM overall was not truly open to the concerns and activities of Oregon's Department of Land Conservation and Development: "I got a little discouraged, and didn't imagine that it would be incorporating any of the land use issues into it." Her discouragement eventually resulted in no longer following up on the program updates. The agency responsible for land use planning in the state known for its long-standing tradition in that field (Kaiser and Godschalk 1995) is no longer involved in the effort to create an ESM. Given the potential impact on land use of the ESM as imagined by its proponents, that's a deeply problematic development.

The final problem associated with the participatory decision-making process in Oregon is more general disengagement. As the experience of a state official reveals, this disengagement does not have to be driven by deep disagreement over the role of place or philosophical opposition to the idea of equivalence but can simply be the result of not being able to communicate effectively across disciplinary boundaries. Yet the cost of the disengagement is real. Oregon's Department of Land Conservation and Development did not write a critical letter to the EPA, nor did it submit a FOIA request to more fully probe if and how the Willamette Partnership and Clean Water Services are trying to "govern through the market." In fact, the Department did the opposite—nothing. Given the problems with follow-through and implementation that have characterized the ESMs described here, this is perhaps an even more serious concern for their proponents. The conclusion to this chapter will review the way in which the various failures, shortcomings and obstacles described reveal the profound challenge that is *participation* in the development of an ESM.

4.e Markets without Participants

The tension between the importance many people attach to the places they know and the need to move (parts) of places around to make an ESM work, as well as to produce compelling evidence for the equivalence between seemingly very different activities, make it difficult to know whom to engage, and when, in the creation of an ESM. This chapter started with the observation that there is no well-established agreement among scholars and/or practitioners on how to develop an ESM at the watershed scale. The sophisticated and inclusive participatory decision-making processes to create ESMs in the three regions analyzed here have met with relatively little success.

Three elements make up the key considerations in designing a participatory process, namely the method used to select participants, the mode(s) of communication during the process and the authority to implement decisions based on the outcome(s) of the process.

The first element, about selection of participants, informs the observation that in all of these cases market enthusiasts made up most of the participants. This has meant that opponents of ESMs typically emerge at a later point in their development, after an agency has issued a permit that includes trading, or when a credit broker makes a trade. Given the broad range of ecosystems and places that ESMs connect, it is not easy to anticipate what kinds of opposition might emerge, or where it will come from geographically. A more serious commitment to identifying and involving a broader set of potential stakeholders in the early stages of ESM development is called for.

The second conclusion about stakeholder participation specific to the development of ESMs is that a strong focus on technology innovation can become an obstacle to inclusiveness and effective communication. Many important assumptions about what to value in an ecosystem are deeply embedded in the design of technical measurement or assessment systems. The inclusion of stakeholders with little ecological or biological training during COTE broadened the range of participants, but made it hard to incorporate their concerns. The models and calculation techniques in the Nutrient Trading Program in the Chesapeake Bay are so complex, as is the institutional context in which they must continue to be adjusted, that the limited number of experts who can follow what is going on can hardly find the time to attend all the relevant meetings and discussions. Non-expert participation is therefore limited, even given the scale and population size of the Chesapeake Bay watershed.

The third and final conclusion about stakeholder engagement in ESM design and implementation is that it is frequently unclear who has the authority required to make a market work. The successes and failures in developing ESMs in the Willamette River basin and Chesapeake Bay watershed shows that a dedicated group of political leaders (not just ESM market enthusiasts) is needed to make meaningful progress. Confusion about the roles and responsibilities of nongovernmental organizations and regulatory agencies can contribute to a lack of legitimacy of the overall effort. The final chapter will take up this challenge of participation in the creation of ESMs and begin to formulate a set of responses to this and the other challenges to ESMs identified in this volume.

Chapter 5

TRADING PLACES

In the movie *Field of Dreams*, a struggling farmer (played by Kevin Costner) repeatedly hears a voice from heaven whisper to him: "If you build it, he will come."[1] The specific instruction is to build a baseball field instead of growing corn, in the middle of Iowa. Once the farmer builds the diamond after an enlightening visit to Fenway Park, the ghosts of famous baseball players start using the field to play ball. As a result, paying audiences flock to this cornfield, alleviating the farmer's financial struggles once and for all. In restoration ecology, this quote is commonly used to describe the expectation that when physical attributes, like water flow or elevated ridges, are restored, biotic responses will rapidly follow, without any additional human intervention (Hilderbrand, Watts and Randle 2005).

It is common to hear proponents of ESMs use this phrase to describe the hope and expectation that, following the creation of a clear set of rules and measurement systems for an ESM, active credit trading would emerge "naturally" and buyers and sellers would "come." In the Willamette River basin, the Chesapeake Bay watershed and the Ohio River basin, this prophecy has not been true when it comes to the creation of ESMs. The three reasons ESMs have failed to materialize are widespread concerns about displacement, the inability to produce equivalence and the problems of participation.

This final chapter revisits those three challenges and develops an argument about their continued relevance as well as a set of ideas on how to learn from the obstacles encountered for environmental policy and planning more generally. The attempts to create integrated markets for multiple ecosystem services at the watershed scale are not isolated incidents, but can be seen as part of a broader ongoing experiment with market-based and market-like instruments in environmental policy and planning. In urban planning, for example, the idea behind transferable development rights (TDR) is somewhat analogous to the logic of ESMs. TDR refers to a practice in which one piece of land in a city

---

1 A review of this movie is available at: http://www.rogerebert.com/reviews/field-of-dreams-1989. Last accessed on 08/27/2015.

is not developed but typically turned into a park or some other public benefit, in return for which the development of another property can exceed some regulation, like restrictions on height or density (Linkous and Chapin 2014). For example, the cost to the community imposed by adding a few stories to a building, in terms of blocked views or increased winds, is compensated by the addition of an amenity like a park or public space. In forestry, the aforementioned sustainability certification program developed by the Forest Stewardship Council is an example of market-oriented experimentation. The efforts to create ESMs in the United States analyzed in this book are meaningfully different from these market-based practices, yet share some of the basic challenges identified here, from concerns over the damage done to a unique place to resistance to the claims made based on complicated measurement systems.

The proponents of ESMs have experienced successes and setbacks in the decade since most of their efforts began. Most of them continue to advocate for and work on the further development of these ESMs, but several have also started to devote significant attention to non-market-based policies and projects.[2] The promise of the ESM idea continues to inspire and influence discussion and debate. One prominent ESM enthusiast published a book-length monograph extolling the virtues of market-based approaches to environmental conservation and restoration, and emphasizing the potential of increased quantification and innovation in this policy domain (Whitworth 2015).

5.a From Displacement to Spatializing Interests

The early success of Clean Water Services' Temperature Management Plan and the Community Tree Planting Challenge helped to convince numerous advocates and policymakers that the time was right to create more ambitious markets for ecosystem services, by increasing their scale and scope. But the apparent simplicity of the shade-trading approach, in which a water treatment plant could fund upstream tree-planting projects to meet its legal obligation to cool water, was not supported by Clean Water Services' experience. This included the development of a detailed tree-planting plan, which incorporated a host of deeply social priorities and considerations, from an emphasis to planting on publicly accessible land to a focus on proximity to urban centers.

The proponents of more holistic, large-scale ESMs did not include such sophisticated spatial planning approaches in their efforts to develop markets

---

2 For example, the Willamette Partnership has started a project on health and nature. More information can be found online: http://willamettepartnership.org/success-stories/health-nature/. Last accessed on 09/10/2015.

and encountered fierce opposition as a result. In Oregon, the National Marine Fisheries Service (NMFS) withdrew its support from the effort to create an ESM for the entire Willamette River basin over concerns that locations of unique importance to the survival of salmonids would be traded away, and replaced by restored habitat sites in far less important locations. Likewise, the City of Portland decided not to pursue participation in the ESM because of its priority to invest in the creation and enhancement of green space within the city limits. In Pennsylvania, the EPA and the Department of Environmental Protection put a halt to the movement of poultry manure from one watershed to the other as a form of credit generation.

The conclusion that place, or more specifically the attachments individuals and organizations express regarding specific socio-spatial configurations, has not been taken into account successfully in the most ambitious attempts to create ESMs in the United States raises two questions. The first is whether it is possible to incorporate the multiple and myriad of preferences, connections and interests related to the places involved in potential ESMs. The second question is, if yes, how?

To seek to answer these questions, it is instructive to return to the original example of shade trading, namely the Clean Water Services permit in the Tualatin River basin in Oregon. The first relevant observation about this program is that the overall area within which credits can be developed is much smaller than the other ESMs described in this book. The Ohio River basin, the Chesapeake Bay watershed and the Willamette River basin are significantly larger than the Tualatin River basin, potentially enabling displacements across great distances. Based on that observation, it seems that the resistance to ESMs based on a discomfort with displacement might diminish if those markets are designed on a relatively small scale, as proposed by well-known Water Quality Trading scholars like Richard Moore (Moore 2014).

However, the basic concerns over displacement expressed by opponents to ESMs did not exclusively focus on distance, but also on the existence of uniquely valuable places and a more general sense of *moving matter out of place*. Such considerations can only be taken into account when they are incorporated into a broadly shared, spatially explicit vision of where displacements can and cannot take place, and what kind(s) of matter can and cannot be moved around. In short, this calls for spatial planning before the investments in restoration and conservation projects commence. The development of the Healthy Streams Initiative performed this function in the Tualatin River basin. This plan included detailed maps of where tree-planting projects should take place, clearly prioritizing publicly accessible land within or close to urban areas in the watershed. The wastewater utility required to reduce its negative environmental impact, Clean Water Services, developed this plan in close collaboration with community groups and other relevant organizations before trading was ever mentioned.

This type of spatial planning severely limits the ability to argue for ESMs based on economic notions of competition and efficiency. Instead, the investment in the creation of ecosystem services by a wastewater treatment utility or electricity company can then be discussed in terms of where and how the implementation of regulatory requirements can benefit communities and ecosystems in a variety of ways. In my view, the appropriate response to the discomfort with displacement encountered in all three major efforts to create integrated ESMs in the United States is to move away from the logic of markets toward a more spatially sensitive logic of collaborative watershed planning. Such planning processes can still be focused on the investments in ecosystem services production necessary to allow specific utilities or companies to meet regulatory requirements more cost-effectively.

The role of the Healthy Streams Initiative in the development of Clean Water Services' shade-trading permit shows that this is not only possible, it is necessary to elicit and incorporate spatial preferences and interests of communities and organizations in the watershed. This type of planning remains exceedingly rare in efforts to create Water Quality Trading schemes or more comprehensive forms of ESMs (Kieser and Logue 2015), yet seems to be the only practical way to deal with the widespread discomfort with displacement inherent to current iterations of the ESM idea.

5.b From Nonequivalence to Improving Places

The creation of a credible system to determine equivalence between adverse environmental impacts and newly created ecosystem services has proven a formidable challenge to the implementation of ESMs. Despite significant investments in modeling and measurement techniques, claims of nonequivalence have emerged in all three efforts to create ESMs analyzed here.

In the Ohio River basin, the differences between new ecosystem services provided by the agricultural sector and pollutants produced by fossil fuel–based electricity generators turned out to be irreconcilable. The Electric Power Research Institute (EPRI) invested time, resources and expertise into the development of a sophisticated model-based system to determine the appropriate amount of phosphorous and nitrogen credits to be awarded for specific Best Management Practices. Much less attention was paid to the activity these credits are intended to compensate for, namely the discharge of water used in the generation of electricity by fossil fuel–burning power plants. It was these power plants that the U.S. EPA regulates as a distinct category. The Effluent Guidelines for the Steam Electric Category issued in 2015, in which the EPA prescribes a set of water quality standards for a range of compounds in discharged water, include nitrogen. These standards cannot be met through trading, but are regarded as a baseline for on-site water treatment that every power plant in this category has to implement. Since the filtration technologies the EPA prescribed in these Effluent Guidelines extract almost all nitrate from the discharged water, power plants will not need to pursue additional nitrogen reductions by purchasing credits. This new guideline has effectively rendered the adverse environmental impact of water discharges by power plants non-equivalent to ecosystem service creation on agricultural lands.

In Oregon, the effort to incorporate wetland banking into an integrated ESM failed in part because the attempt to more precisely measure the equivalence between a destroyed wetland and a restored wetland faltered. While the Oregon Rapid Wetland Assessment Protocol (ORWAP) could count on support from experts and agency staff, the organizations actually responsible for developing and restoring wetlands have decided to continue using alternative wetland assessment procedures. Since the regulatory regime for wetland banking has developed over multiple decades, the preference to stick with the existing procedures perhaps should not have come as a surprise. The measurement systems developed and used in (proposed) markets for ecosystem services all face this challenge, since most federal environmental regulations that could force permit holders to actually buy credits are decades old. This long history of implementation means that measurement and assessment practices are typically well established and hard to destabilize. The promise that the creation of new measurement systems for ESMs will result in more precise ways of achieving the goals of the Clean Water and Endangered Species Acts has done little or nothing to persuade all relevant stakeholders to actually participate in these markets. In fact, in an area with a relatively short history of implementation of the Endangered Species Act, namely habitat banking for Blue Fender's Butterfly, the proponents of ESMs were unable to convince key stakeholders of the efficacy of a newly created measurement system.

The limited success of these ambitious and well-resourced efforts to create measurement systems that can show the equivalence between polluting activities and the provision of ecosystem services raises the question if it is possible at all to show that certain (parts of) ecosystems provide specific quantities of services that are meaningfully equivalent to human activities in other locations. To answer that question, it is instructive to return to the shade-trading example in the Tualatin River basin that inspired so many of the later efforts to create ESMs elsewhere, at larger scales and for broader suites of services.

In the Tualatin, the "polluter" or credit buyer, namely the water utility Clean Water Services, never argued that the tree-planting projects would exclusively and precisely compensate for the elevated water temperature in its discharge. Instead, Clean Water Services effectively committed to embark on a broader effort to improve the overall water quality in the region by purchasing water rights and engaging in a broader environmental restoration program in the region that combined volunteer efforts and financial incentives. Clean Water Services described this restoration program in much detail in the organization's Healthy Streams Initiative, the product of an extensive participatory planning process itself. Its argument about the equivalence between adverse impact, namely the elevated water temperature, and the compensatory actions relied on much more than simply the Shade-a-Lator, despite the catchy name and relatively straightforward implementation of that measurement system.

Yet that very narrow measurement system has captured the imagination of many of the ESM proponents in other parts of the country. This reduction of the problem of (non-) equivalence has resulted in a search for the magic measurement system that can simultaneously meet the very divergent logics of state, market and science. This of course is a challenge that geographer Morgan Robertson (2006; 2007; 2012) has repeatedly shown to be practically insurmountable and theoretically problematic as well.

A different reading of the Tualatin River basin example provides insight into how equivalence can be argued more convincingly when a much broader range of environmental impacts and issues is explicitly and thoroughly taken into consideration. Rather than emphasizing how the water treatment utility was pursuing trading as a low-cost permit compliance strategy, Clean Water Services collaboratively developed a watershed-wide environmental improvement plan, in which it would invest significant amounts of money. The measurement system that could support one of those elements, the Shade-a-Lator, was only a part of that much more encompassing vision of what Clean Water Services should do to compensate for its adverse environmental impacts.

The production of equivalence is not purely a matter of measurement and modeling, but relies on a much broader set of arguments and ideas about what appropriate strategies are to compensate for damage to the environment. If the proponents of investments in ecosystem service provision want to follow the example of the Tualatin River basin case, they could focus less on detailed measurement systems and more on the development of broadly supported plans to channel mandatory investments in pollution reduction into much more integrated plans to improve environmental quality in watersheds as a whole. That requires not only a shift in emphasis by ESM proponents away from precise measurement; it also suggests a fundamentally different role for the groups that oppose trading-based strategies.

5.c From Opposition to Building Consensus

The process of seeking to create an ESM inevitably involves many organizations and individuals. Using three common elements of participatory decision-making processes to analyze the efforts to create ESMs in the Chesapeake Bay, the Ohio River and the Willamette River watersheds reveals formidable barriers to successful participation of those stakeholders. While the proponents of ESMs have used different types of participatory processes in their efforts in these three regions, the outcomes have been remarkably similar. First, in all three areas, opponents have emerged and continued to oppose the creation of these new institutions, regardless of the type and level of their involvement in the ESM development process. Second, the appropriate relationship between the proponents of ESMs and responsible regulatory agencies in the design and development of these institutions remains contested, even among some of the market enthusiasts. Third, the extent to which advocacy organizations and the public at large should be able to continuously scrutinize the specific impacts of ecosystem service credit-generating projects is controversial in all regions. While these obstacles to meaningful and successful participation seem formidable, the three market-making efforts also reveal potential opportunities to overcome them.

The development of an integrated Water Quality Trading scheme for the entire Chesapeake Bay watershed seems to be somewhat disjointed and most closely follows a traditional regulatory implementation process. The World Resources Institute clearly plays an important role in bringing together stakeholders in various states and by providing technical expertise related to Nutrient Net. However, it is the U.S. EPA which has the responsibility to ensure that actions of the individual states in implementing the regulatory requirements flowing from the Chesapeake Bay TMDL amount to an overall improvement of water quality across the watershed. The simultaneous and often overlapping efforts to create or enhance trading programs in individual states, by both WRI and the EPA, combined with these organizations' goals to create opportunities for credit trading across state boundaries, has resulted in a bewildering array of working groups and steering committees, in which only the most dedicated experts and officials continue to take part. This suggests that a more focused effort to engage possible participants in conversations about where and how to invest in ecosystem service conservation and restoration might yield a greater chance of success.

In the Ohio River basin, a more focused effort has taken place over the past 10 years, initiated and led by the Electric Power Research Institute. The structure of this participatory process has been sophisticated yet clear. Participants in this process have represented a broad range of organizations and agencies, including at least one that has long been publicly opposed to credit trading–based environmental policy instruments, namely the Sierra Club. Despite this more inclusive and targeted participation process, resistance persists and questions about the schedule of implementation and oversight continue to go unanswered. When EPRI decided to focus its effort to create an ESM in the

Ohio River basin, the organization touted as one of the benefits of participating early on in the development of an ESM that regulated entities could
influence the eventual content of those requirements. Organizations like the
Sierra Club were highly critical of that line of argumentation, and expressed
as much to the U.S. EPA. A more direct challenge to the implementation of
the ESM as envisioned by EPRI is the question of who maintains oversight of
credit generating projects upon and following installation. In EPRI's view, the
Soil and Water Conservation Districts perform this critical role in collaboration with the responsible state agency, but several environmental advocacy
organizations believe a far broader range of interested entities ought to be
able to check those projects. The absence of a clear regulatory requirement
that would incentivize power plants in the Ohio River basin to meaningfully
reduce nitrogen levels through the acquisition of credits means that this concern remains hypothetical. After the electricity companies purchased the first
round of credits on an explicitly voluntary basis, no more credits have been
offered for sale. The ongoing uncertainty over the numeric nutrient criteria
and the new requirement to upgrade on-site filtration technologies as a result
of the updated technology-based Effluent Guidelines has made any large-
scale investments in nitrogen credits by power plants in the near future highly
unlikely. These experiences in the Ohio River basin highlight the urgent call
expressed by prominent environmental organizations to be able to independently assess, if at least through the provision of spatially explicit information,
the environmental restoration activities that are generating credits throughout
the lifespan of the ESM.

The third large-scale effort to create an ESM, the Willamette Partnership's
Counting on the Environment process, exemplified many of the best practices
outlined by experts on collaborative environmental policymaking. Somewhat
surprisingly, given the emphasis on neutral facilitation and broad participation, the designers of this process excluded known opponents of market-based
environmental policy tools. This decision did not prevent those groups from
forcefully voicing their opposition to certain elements of the proposed ESM,
both through open records requests and direct appeals to senior U.S. EPA officials. A secondary challenge to the participatory ESM development process
in Oregon turned out to be the inability to effectively communicate across
disciplinary boundaries. As many of the discussion during the process were
framed in explicitly ecological terms, representatives of organizations and
agencies more focused on the human aspect of land use and development felt
somewhat excluded from the decision making. This resulted in a gradual with-
drawal from the overall project by stakeholders that, based on their expertise
and regulatory role, might have been central players. This analysis of participation in the attempt to develop an ESM in Oregon reveals that organizations

require active engagement to stay involved, which may necessitate a move away from a focus on ecosystem services in an ecological sense, and toward a broader discussion about socio-ecological systems and the particular land-scape features in locations in the area of interest. Of course, at the scale of trading the proponents of ESMs envision, it would be exceedingly compli-cated and resource-intensive to engage relevant representatives in those kinds of negotiations.

Organizing and managing the participation of an inclusive group of stake-holders in the development and implementation of an ESM turns out to be a formidable challenge. Upon closer inspection of the specific moments and issues around which resistance emerges or disengagement occurs, the inter-ventions that might help proponents overcome those problems move away from the appealing imagination of an efficient, large-scale market that pro-vides cost-effective environmental benefits. Instead, the regulatory-driven pro-cess that enables all types of groups ongoing opportunities to analyze and discuss environmental restoration strategies and that incorporates disciplinary perspectives well beyond ecology begins to sound much more like a conven-tional landscape-scale conservation and restoration planning effort. It is this observation that the concluding section will focus on.

5.d Conclusion

The challenges that have persistently emerged in the efforts to create markets for ecosystem services are multiple and deeply connected to some of the cen-tral imaginations of what makes those markets desirable in the first place. The hoped-for efficiencies, cost reductions and precisely quantified environmental improvements have not yet taken shape, despite a decade of work by a com-mitted, well-funded network of professionals and experts in three regions of the United States. So while the promise of the ESM might persist, the practi-cal experiences reveal that the Clean Water Services program in the Tualatin River basin remains a rarity, an inspiring example to many but a tough act to follow. Perhaps the central focus on the market-like aspects of this example has actually been a disservice to those inspired by its success. The story of the investments in water cooling and shade creation in the Tualatin River basin can just as easily and perhaps more accurately be hailed as a successful example of collaborative watershed planning. Yet this is not how that story is typically recounted.

The problems of displacement, nonequivalence and participation might be overcome if the proponents described in this book can relax or even abandon some of the key tenets of the very ESM idea. The kinds of pragmatic adjust-ments sketched in the preceding paragraphs suggest a project more akin to landscape-scale conservation and restoration planning than market making or credit trading. In such a planning process, participants will be able to more easily advocate for the connections they feel to particular places, make the determination of what kinds of restoration activities are meaningfully equivalent to harmful impacts and scrutinize the actual implementation of any kind of ecosystem changes or preservation actions. It falls outside of the scope of this book to describe the history of watershed and landscape planning in detail, but it's relevant to note that those are vibrant areas of scholarship and practice.

The future of ESMs in the United States, and, more specifically, the kind of ambitious large-scale efforts to create integrated institutions in which multiple credit types can be bought and sold in single transactions, is highly uncertain. The attempts described in this book all seem to have stalled in one form or another, and no new efforts that are similar in scope and scale have emerged in recent years. In some ways, the market enthusiasts at the center of this project seem to have realized some of these challenges and responded by narrowing and focusing their efforts.

Yet, given the real threats to environmental, economic and social well-being, it is clearly necessary to continue to work to enhance and restore places and spaces, find creative ways to reduce and compensate for adverse environmental impacts and engage with one another to build consensus about where and how to focus those efforts. It is these goals that drive much of the hope and enthusiasm for markets for ecosystem services. As some of that effervescence has waned, it remains as important as ever to maintain a commitment to pursuing those broader goals.

LIST OF INTERVIEWS

Organization	Name	Date of Interview
Adamus Resource Assessment	Paul Adamus	July 15, 2011
American Electric Power	Tim Lohner	March 26, 2015
American Farmland Trust	Brian Brandt	January 29, 2015
Alliance for the Chesapeake Bay	Al Todd	November 8, 2011
Bonneville Environmental Foundation	Kendra Smith	July 25, 2011
City of Eugene	Eric Wold	August 18, 2011
Clean Water Services	Bruce Roll	July 29, 2011
Columbiana Soil and Water Conservation District	Pete Conkle	January 22, 2015
Defenders of Wildlife	Sara Vickerman	June 12, 2011
Electric Power Research Institute	Jessica Fox	January 29, 2015
Environmental Banc and Exchange	George Kelly	November 20, 2013
Environmental Law and Policy Center	Brad Klein	March 27, 2015
Forest Trends/University of Maryland	Dan Nees	November 14, 2011
GreenVest	Doug Lashley	November 14, 2011
Guillozet Consulting	Peter Guillozet	July 12, 2011
Institute for Natural Resources	Jimmy Kagan	August 15, 2011
Iowa Farm Bureau Federation	Rick Robinson	March 6, 2015
Kearns and West	Deb Nudelman	July 22, 2011
Kentucky Department of Natural Resources	Jamie Ponder	February 17, 2015
Madsen Environmental (then USDA Forest Service)	Becca Madsen	January 22, 2013

Organization	Name	Date of Interview
Maryland Department of Agriculture	Susan Payne	November 9, 2011
Meyer Memorial Trust	Pam Wiley	July 26, 2011
Miami Conservancy District	Sarah Hippensteel-Hall	March 10, 2015
Mid-Ohio Regional Planning Commission	David Rutter	January 22, 2015
National Oceanic and Atmospheric Administration	Marc Liverman	July 26, 2011
Northwest Environmental Advocates	Nina Bell	April 30, 2015
Ohio Department of Natural Resources	Dorothy Farris	March 12, 2015
Ohio Environmental Protection Agency	Dan Dudley	March 17, 2015
Ohio Environmental Protection Agency	Gary Stuhlfauth	March 24, 2015
Ohio Farm Bureau	Larry Antosch	February 17, 2015
Ohio River Valley Water Sanitation Commission	Peter Tennant	March 26, 2015
Oregon Consensus	Turner Odell	August 4, 2011
Oregon Dept. of State Lands	Dana Hicks	July 18, 2011
Oregon Dept. of Agriculture	Dave Wilkinson	August 3, 2011
Oregon Dept. of Environmental Quality	Rainei Nomura	May 27, 2011
Oregon Dept. of Environmental Quality	Ryan Michie	May 27, 2011
Oregon Dept. of Environmental Quality	Eugene Foster	May 27, 2011
Oregon Dept. of Forestry	Jeff Brandt	July 27, 2011
Oregon Dept. of Forestry	Mike Cafferata	July 27, 2011
Oregon Dept. of Land Conservation and Development	Katherine Daniels	July 18, 2011
Oregon Watershed Enhancement Board	Renee Davis-Born	July 20, 2011
Parametrix	Kevin Halsey	June 19, 2011
Penn Futures	Alissa Burger	March 1, 2013
Pennsylvania Department of Environmental Protection	Ann Roda	November 17, 2013

Organization	Name	Date of Interview
Pinchot Institute	Eric Sprague	November 17, 2013
Portland Bureau of Environmental Services	Mike Reed	August 5, 2011
Portland Bureau of Healthy Working Rivers	Heidi Berg	May 26, 2011
Portland Bureau of Healthy Working Rivers	Rick Bastasch	May 26, 2011
Portland Bureau of Healthy Working Rivers	Ann Beier	May 26, 2011
Red Barn Consulting	Peter Hughes	January 30, 2013
Sierra Club—Hoosier Chapter	Bowden Quinn	April 2, 2015
The Freshwater Trust	Scott Peerman	July 21, 2011
The Freshwater Trust	David Primozich	July 14, 2011
U.S. Army Corps of Engineers	Jaimee Davis	July 13, 2011
U.S. Army Corps of Engineers	Bill Abadie	July 13, 2011
U.S. Department of Agriculture	Carl Lucero	November 10, 2011
U.S. Environmental Protection Agency	Kevin DeBell	November 15, 2011
U.S. Environmental Protection Agency	Ellen Gilinsky	April 6, 2015
U.S. Environmental Protection Agency	Bob Rose	April 6, 2015
U.S. Environmental Protection Agency	Yvonne Vallette	November 7, 2011
U.S. Environmental Protection Agency	Matt Weber	July 20, 2011
U.S. Forest Service	Robert Deal	July 21, 2011
U.S.D.A. Natural Resources Conservation Service	Anthony Nott	March 23, 2015
U.S.D.A. Natural Resources Conservation Service	Meta Lofstgaarden	July 22, 2011
University of Oregon	David Hulse	August 19, 2011
Watershed Stewardship, Inc.	Tom Simpson	November 10, 2011
Willamette Partnership	Bobby Cochran	July 8, 2011
Willamette Partnership	Devin Judge-Lord	May 27, 2010
World Resources Institute	Todd Gartner	March 4, 2013
World Resources Institute	Mindy Selman	November 15, 2011

BIBLIOGRAPHY

Abdalla, C. 2008. Land Use Policy: Lessons from Water Quality Markets. *Choices*, *23*(4).

Achterman, G. L., & Mauger, R. 2010. The State and Regional Role in Developing Ecosystem Service Markets. *Duke Envtl. L. & Pol'y F, 20*, 291–417.

Ackerman, B. A., & Stewart, R. B. 1987. Reforming Environmental Law: The Democratic Case for Market Incentives. *Colum. J. Envtl. L., 13*, 171.

Adamus, P. R., & Stockwell, L. T. 1983. "A Method for Wetland Functional Assessment: Volume 1 Critical Review and Evaluation Concepts." Washington, DC: Federal Highway Administration. http://trid.trb.org/view.aspx?id=205071.

Adamus, P. R., Morlan, J., & Verble, K. 2009a. *Calculator Spreadsheet, Databases, and Data Forms for the Oregon Rapid Wetland Assessment Protocol (ORWAP).* (version 2.0). Salem: Oregon Dept. of State Lands.

Adamus, P. R., Morlan, J., & Verble, K. 2009b. "Manual for the Oregon Rapid Wetland Assessment Protocol (ORWAP). Version 2.0." Salem: Oregon Dept. of State Lands.

Alexander, R. B., Smith, R. A., Schwarz, G. E., Boyer, E. W., Nolan, J. V., & Brakebill, J. W. 2008. "Differences in Phosphorus and Nitrogen Delivery to the Gulf of Mexico from the Mississippi River Basin." *Environmental Science & Technology 42*(3): 822–830.

Anderson, T. L., & Leal, D. 1991. *Free Market Environmentalism.* San Francisco, CA; Boulder, CO: Pacific Research Institute for Public Policy; Westview Press.

Appadurai, A. 1996. *Modernity at Large: Cultural Dimensions of Globalization.* Minneapolis: University of Minnesota Press.

Arestis, P., & Sawyer, M. C. 2004. *The Rise of the Market: Critical Essays on the Political Economy of Neo-liberalism.* Cheltenham, UK and Northampton, MA: Edward Elgar.

Argyris, C., & Schön, D. A. 1978. *Organizational Learning.* Reading, MA: Addison-Wesley.

Armsworth, P. R., Chan, K. M. A., Daily, G. C., Ehrlich, P. R., Kremen, C. T., Ricketts, T. H., & Sanjayan, M. A. 2007. "Ecosystem-Service Science and the Way Forward for Conservation." *Conservation Biology 21*(6): 1383–1384.

Bardach, E. 1977. *The Implementation Game: What Happens after a Bill Becomes a Law.* Cambridge, MA: MIT Press.

Bean, M., Bonnie, R., Male, T., & Searchinger, T. 2003. "The Private Lands Opportunity: The Case for Conservation Incentives." Environmental Defense. http://www.fws.gov/southeast/grants/pdf/2677_ccireport.pdf.

Beck, U. 1992. *Risk Society: Towards a New Modernity.* London; Newbury Park, CA: Sage.

BenDor, T., & Brozović, N. 2007. Determinants of Spatial and Temporal Patterns in Compensatory Wetland Mitigation. *Environmental Management, 40*(3), 349–364.

BenDor, T., Brozović, N., & Pallathucheril, V. G. 2007. Assessing the Socioeconomic Impacts of Wetland Mitigation in the Chicago Region. *Journal of the American Planning Association, 73*(3), 263–282.

Bennett, G. & Carroll, N. 2014. *Gaining Depth: State of Watershed Payments 2014.* Washington DC: Forest Trends.

Bennett, G., Carroll, N., & Hamilton, K. 2013. "Charting New Waters: State of Watershed Payments 2012." Washington, DC: Forest Trends. http://www.forest-trends.org/documents/files/doc_3308.pdf.

Benson, D., Jordan, A., Cook, H., & Smith, L. 2013. Collaborative Environmental Governance: Are Watershed Partnerships Swimming or Are They Sinking? *Land Use Policy*, *30*(1), 748–757.

Bidwell, R. D., & Ryan, C. M. 2006. Collaborative Partnership Design: The Implications of Organizational Affiliation for Watershed Partnerships. *Society & Natural Resources*, *19*(9), 827–843.

Bingham, G., Bishop, R., Brody, M., Bromley, D., Clark, E. (Toby), Cooper, W., ... Suter, G. 1995. Issues in Ecosystem Valuation: Improving Information for Decision Making. *Ecological Economics*, *14*(2), 73–90.

Blankenship, K. 2005, June. Solutions Sought for Excess Manure Piling Up on Farms. *Chesapeake Bay Journal*. Retrieved from http://www.bayjournal.com/article/solutions_sought_for_excess_manure_piling_up_on_farms.

Bockstael, N., Costanza, R., Strand, I., Boynton, W., Bell, K., & Wainger, L. 1995. Ecological Economic Modeling and Valuation of Ecosystems. *Ecological Economics*, *14*(2), 143–159.

Bourdieu, P. 1977. *Outline of a Theory of Practice*. Cambridge, UK; New York: Cambridge University Press.

Bourdieu, P., & Wacquant, L. J. D. 1992. *An Invitation to Reflexive Sociology*. Chicago: University of Chicago Press.

Boyd, J., & Banzhaf, S. 2007. What Are Ecosystem Services? The Need for Standardized Environmental Accounting Units. *Ecological Economics*, *63*(2–3), 616–626.

Boyd, M. S. 1996. "Heat Source: Stream, River and Open Channel Temperature Prediction." Master's thesis, Corvallis: Oregon State University. http://ir.library.oregonstate.edu/xmlui/handle/1957/27036.

Branosky, E., Jones, C., & Selman, M. 2011, May. Comparison Tables of State Nutrient Trading Programs in the Chesapeake Bay Watershed. World Resources Institute Fact Sheet Version 1. http://www.wri.org/sites/default/files/comparison_tables_of_state_chesapeake_bay_nutrient_trading_programs.pdf.

Brown, S. 2011. "Navigating the Edges: An Examination of the Relationship between Boundary Spanning, Social Learning, and Partnership Capacity in Water Resource Management." Unpublished dissertation. Portland, OR: Portland State University.

Callon, M. 1986. "Some Elements of a Sociology of Translation: The Domestication of the Scallops and St Brieuc Fishermen." In *Power, Action and Belief*. Sociological Review Monograph 32. 196–233. London, UK: Routledge and Kegan Paul.

Carpenter, S. L., & Kennedy, W. J. D. 1988. *Managing Public Disputes: A Practical Guide to Handling Conflict and Reaching Agreements*. San Francisco, CA: Josey-Bass.

Carroll, N., Fox, J., & Bayon, R. 2008. *Conservation and Biodiversity Banking: A Guide to Setting Up and Running Biodiversity Credit Trading Systems*. Earthscan.

Carson, Lyn. 2008. "The IAP2 Spectrum: Larry Susskind in Conversation with IAP2 Members." *International Journal of Public Participation* *2*(2), 67–84.

Casey, F., Vickerman, S., Hummon, C., & Taylor, B. 2006. "Incentives for Biodiversity Conservation: An Ecological and Economic Assessment." Washington, DC: Defenders of Wildlife.

Cash, D. W., Clark, W. C., Alcock, F., Dickson, N. M., Eckley, N., Guston, D. H., Jäger, J., & Mitchell, R. B. 2003. "Knowledge Systems for Sustainable Development." *Proceedings of the National Academy of Sciences 100*(14): 8086–8091.

Cashore, B. W., Auld, G., & Newsom, D. 2004. *Governing through Markets: Forest Certification and the Emergence of Non-state Authority*. New Haven, CT: Yale University Press.

Castree, N. 2008. Neoliberalising Nature: The Logics of Deregulation and Reregulation. *Environment and Planning A, 40*(1), 131–152.

Century Engineering. 2011. "Nutrient Trading Evaluation Report." 113005.00. Harrisburg, PA: Penn Future. http://www.pennfuture.org/UserFiles/File/Water/Resp-Farm/Report_NutrientTradingEval_20110919.pdf.

Chichilnisky, G., & Heal, G. M. 2000. *Environmental Markets: Equity and Efficiency*. New York: Columbia University Press.

Chomsky, N. 1999. *Profit over People: Neoliberalism and the Global Order*. New York, Toronto, London: Seven Stories Press.

Coase, R. H. 1937. "The Nature of the Firm." *Economica*: New Series *4*(16) (November): 386–405.

Cochran, J. R. 2008. "Giving Process Its Due: Can Collaboration Help Environmental Markets Succeed?" Portland, OR: Portland State University. Unpublished dissertation Available at: http://gradworks.umi.com/33/43/3343759.html.

Cochran, J. R., & Logue, C. 2011. "A Watershed Approach to Improve Water Quality: Case Study of Clean Water Services' Tualatin River Program." *JAWRA Journal of the American Water Resources Association 47*(1): 29–38.

Cochran, J. R., & Robinson Maness, N. 2011. "Measuring Up: Synchronizing Biodiversity Measurement Systems for Markets and Other Incentive Programs." U.S. Department of Agriculture Office of Environmental Markets. Hillsboro, OR.

Conniff, R. 2009. "The Political History of Cap and Trade." *Smithsonian Magazine*, August. http://www.smithsonianmag.com/science-nature/Presence-of-Mind-Blue-Sky-Thinking.html.

Cooley, D., & Olander, L. 2011. "Stacking Ecosystem Services Payments: Risks and Solutions." Working Paper NI WP 11-04. Nicholas Institute for Environmental Policy Solutions. Duke University—Nicholas Institute.

Corbera, E., Brown, K., & Adger, W. N. 2007. The Equity and Legitimacy of Markets for Ecosystem Services. *Development & Change, 38*(4), 587–613.

Corburn, J. 2005. *Street Science: Community Knowledge and Environmental Health Justice*. Cambridge, MA: MIT Press.

Cornell, S. 2011. The Rise and Rise of Ecosystem Services: Is "Value" the Best Bridging Concept between Society and the Natural World? *Procedia Environmental Sciences, 6*, 88–95.

Costanza, R., d'Arge, R., Groot, R. de, Farber, S., Grasso, M., Hannon, B., & Belt, M. van den. 1997. The Value of the World's Ecosystem Services and Natural Capital. *Nature, 387*, 253–260.

Cowardin, L. M., Carter, V., Golet, F., & LaRoe, E. 1979. "Classification of Wetlands and Deepwater Habitats of the United States." FWS/OBS-79/31. Washington, DC: U.S. Department of the Interior Fish and Wildlife Service. http://www.npwrc.usgs.gov/resource/wetlands/classwet/.

Cronon, W. 1983. *Changes in the Land: Indians, Colonists, and the Ecology of New England*. New York: Macmillan.

Cronon, W. 1996. *Uncommon Ground: Rethinking the Human Place in Nature*. New York: W. W. Norton & Company.

Cyert, R., & March, J. 1963. *Behavioral Theory of the Firm*. Englewood Cliffs, NJ: Prentice-Hall.

Dales, J. H. 1968. *Pollution, Property & Prices.* Toronto: University of Toronto Press.

Daily, G. C. (ed.) 1997. *Nature's Services: Societal Dependence on Natural Ecosystems.* New York: Island Press.

Daily, G. C., & Matson, P. A. 2008. Ecosystem Services: From Theory to Implementation. *Proceedings of the National Academy of Sciences, 105*(28), 9455–9456.

Daily, G. C., Polasky, S., Goldstein, J., Kareiva, P. M., Mooney, H. A., Pejchar, L., Ricketts, T. H., Salzman, J., & Shallenberger, R. 2009. Ecosystem Services in Decision Making: Time to Deliver. *Frontiers in Ecology and the Environment, 7*(1) (February 1), 21–28.

Dauer, D. M., Ranasinghe, J. A., & Weisberg, S. B. 2000. Relationships Between Benthic Community Condition, Water Quality, Sediment Quality, Nutrient Loads, and Land Use Patterns in Chesapeake Bay. *Estuaries, 23*(1) (February), 80–96.

De Groot, R. S., Wilson, M. A., & Boumans, R. M. 2002. A Typology for the Classification, Description and Valuation of Ecosystem Functions, Goods and Services. *Ecological Economics, 41*(3), 393–408.

Dempsey, J., & Robertson, M. M. 2012. Ecosystem Services Tensions, Impurities, and Points of Engagement Within Neoliberalism. *Progress in Human Geography, 36*(6) (December 1), 758–779.

Dennison, M. 1997. *Wetland Mitigation: Mitigation Banking and Other Strategies for Development and Compliance.* Rockville, MD: Government Institutes.

Devall, B., & Sessions, G. 1985. *Deep Ecology.* Salt Lake City, UT: Gibbs Smith.

Dewey, J. 1963. *Experience and Education.* New York: Collier Books.

Diaz, D., Hamilton, K., & Johnson, E. 2011. *State of the Forest Carbon Markets 2011: From Canopy to Currency.* Ecosystem Marketplace. Retrieved from http://www.forest-trends.org/documents/files/doc_2963.pdf. Last accessed on July 17, 2013.

Donahue, J. D. 1989. *The Privatization Decision: Public Ends, Private Means.* New York: Basic Books.

Donigian, A. S., Crawford, N. H., U.S. Environmental Protection Agency Office of Research and Development, & Hydrocomp Inc. 1976. *Modeling Nonpoint Pollution from the Land Surface.* U.S. Environmental Protection Agency, Office of Research and Development, Environmental Research Laboratory.

Donlan, J. 2005. Re-wilding North America. *Nature, 436*(7053) (August 18), 913–914.

Dooling, S. 2009. Ecological Gentrification: A Research Agenda Exploring Justice in the City. *International Journal of Urban and Regional Research, 33*(3), 621–639.

Douglas, M. 1966. *Purity and Danger: An Analysis of Concept of Pollution and Taboo.* London: Routledge & Paul.

Doyle, M. W., & Yates, A. J. 2010. Stream Ecosystem Service Markets under No-Net-Loss Regulation. *Ecological Economics, 69*(4), 820–827.

Duijn, M. 2009. Embedded Reflection on Public Policy Innovation: A Relativist/Pragmatist Inquiry into the Practice of Innovation and Knowledge Transfer in the WaterINNovation Program. Delft: Eburon.

Dunleavy, P., & Hood, C. 1994. From Old Public Administration to New Public Management. *Public Money & Management, 14*(3), 9–16.

Effinger, A. 2011, March 31. Harry & David's Bankruptcy Rocks Medford, Oregon. *BusinessWeek: Magazine.* Retrieved from http://www.businessweek.com/magazine/content/11_15/b4223051566568.htm.

Engel, S., Pagiola, S., & Wunder, S. 2008. Designing Payments for Environmental Services in Theory and Practice: An Overview of the Issues. *Ecological Economics, 65*(4), 663–674.

ELI. 2007. *Mitigation of Impacts to Fish and Wildlife Habitat: Estimating Costs and Identifying Opportunities.* Washington, DC: Environmental Law Institute.

Ellerman, A. D. 2000. *Markets for Clean Air: The U.S. Acid Rain Program.* Cambridge, UK; New York: Cambridge University Press.

EPRI. 2002. "Water Quality Trading Guidance Manual: An Overview of Program Design Issues and Options.pdf." Technical 1005179. Palo Alto, CA: EPRI. http://www.epri.com/abstracts/Pages/ProductAbstract.aspx?ProductId=000000000001005179.

EPRI. 2006. "Water Quality Trading Opportunities for Electric Power Companies." Technical 1013193. Program on Technology Innovation. Palo Alto, CA: EPRI. http://www.epri.com/abstracts/Pages/ProductAbstract.aspx?ProductId=00000000000 1013193.

EPRI. 2007a. "Water Quality Trading Program for Nitrogen." Technical 1014646. Program on Technology Innovation. Palo Alto, CA: EPRI. http://www.epri.com/abstracts/Pages/ProductAbstract.aspx?ProductId=000000000001014646.

EPRI. 2007b. "Program on Technology Innovation: Water Quality Trading Pilot Programs." 1015409. Palo Alto, CA: EPRI.

EPRI. 2010. "Program on Technology Innovation: Ohio River Water Quality Trading Pilot Program—Business Case for Power Company Participation, 2008." 1018861. Palo Alto, CA: EPRI.

EPRI. 2011. "Use of Models to Reduce Uncertainty and Improve Ecological Effectiveness of Water Quality Trading Programs." Technical 000000000001023610. Palo Alto, CA: EPRI. http://www.epri.com/abstracts/Pages/ProductAbstract.aspx?ProductId=000000000001023610.

EPRI. 2013a. "Developing Greenhouse Gas Emissions Offsets by Reducing Nitrous Oxide Emissions in Agricultural Crop Production." 1023669. Palo Alto, CA: EPRI.

EPRI. 2013b. "EPRI Comments on Proposed Effluent Limitations Guidelines Rule." 3002002231. Palo Alto, CA: EPRI. http://www.epri.com/Pages/EPRI-Comments-on-Proposed-Effluent-Limitations-Guidelines-Rule.aspx.

EPRI. 2013c. "Case Studies of Water Quality Trading Being Used for Compliance with Nutrient NPDES Permit Limits.pdf." Technical 000000003002001454. Palo Alto, CA: EPRI. http://www.epri.com/abstracts/Pages/ProductAbstract.aspx?ProductId=00 0000003002001454.

EPRI. 2014. "WARMF Watershed Modeling for Nutrient Trading in the Ohio River Basin: Analysis of Upper Ohio, Middle Ohio, Great Miami, Muskingum, and Scioto River." Technical Report 3002002811. Palo Alto, CA: EPRI.

Estrada, M., Corbera, E., & Brown, K. 2008. "How Do Regulated and Voluntary Carbon-Offset Schemes Compare?" Tyndall Centre for Climate Change. Retrieved from http://files.uniteddiversity.com/Climate_Change/How_do_regulated_and_voluntary_carbon-offset_schemes_compare.pdf. Last accessed on 07/17/2013.

Faeth, P. 2000. *Fertile Ground: Nutrient Trading's Potential to Cost-effectively Improve Water Quality.* Washington, DC: World Resources Institute.

Farley, J., & Costanza, R. 2010. Payments for Ecosystem Services: From Local to Global. *Ecological Economics, 69*(11) (September 15), 2060–2068.

Faust, D. G. 2008. *This Republic of Suffering: Death and the American Civil War.* New York: Alfred A. Knopf.

Fischer, F., & Forester, J. 1993. *The Argumentative Turn in Policy Analysis and Planning.* Durham, NC: Duke University Press.

Fisher, B., Turner, R. K., & Morling, P. 2009. Defining and Classifying Ecosystem Services for Decision Making. *Ecological Economics, 68*(3), 643–653.

Floress, K., Prokopy, L. S., & Allred, S. B. 2011. It's Who You Know: Social Capital, Social Networks, and Watershed Groups. *Society & Natural Resources, 24*(9), 871–886.

Forester, J. 2009. *Dealing with Differences: Dramas of Mediating Public Disputes*. Oxford; New York: Oxford University Press.

Fox, J. 2011. "U.S. National Opinion Survey on Stacking Environmental Credits: Definition, Status, and Predictions of Wetland, Species, Carbon and Water Quality Credit Stacking." Palo Alto, CA: Electric Power Research Institute.

Fox, J., Gardner, R., & Maki, T. 2011. Stacking Opportunities and Risks in Environmental Credit Markets. *Environmental Law Reporter, 41*(2) (February), 10121–10125.

Fox, J., & Nino-Murcia, A. 2005. Status of Species Conservation Banking in the United States. *Conservation Biology, 19*(4), 996–1007.

Freeman, J., & Kolstad, C. D. 2007. *Moving to Markets in Environmental Regulation: Lessons from Twenty Years of Experience*. Oxford; New York: Oxford University Press.

Frumkin, H. 2003. Healthy Places: Exploring the Evidence. *American Journal of Public Health, 93*(9) (September), 1451–1456.

Fung, A. 2006. Varieties of Participation in Complex Governance. *Public Administration Review, 66*, 66–75.

Fung, A., Wright, E. O., & Abers, R. 2003. *Deepening Democracy: Institutional Innovations in Empowered Participatory Governance*. London: Verso.

García-Oliva, F., & Masera, O. R. 2004. Assessment and Measurement Issues Related to Soil Carbon Sequestration in Land-Use, Land-Use Change, and Forestry (LULUCF) Projects Under the Kyoto Protocol. *Climatic Change, 65*(3) (August 1), 347–364.

Gardner, R. & Fox, J. 2013. "The Legal Status of Environmental Credit Stacking." Stetson College of Law Legal Studies Research Paper, 2014-2.

Glenna, L. L. 2010. Value-Laden Technocratic Management and Environmental Conflicts: The Case of the New York City Watershed Controversy. *Science, Technology and Human Values, 35*(1), 81–112.

Gómez-Baggethun, E., De Groot, R., Lomas, P. L., & Montes, C. 2010. The History of Ecosystem Services in Economic Theory and Practice: From Early Notions to Markets and Payment Schemes. *Ecological Economics, 69*(6), 1209–1218.

Gössling, S., Broderick, J., Upham, P., Ceron, J.-P., Dubois, G., Peeters, P., & Strasdas, W. 2007. Voluntary Carbon Offsetting Schemes for Aviation: Efficiency, Credibility and Sustainable Tourism. *Journal of Sustainable Tourism, 15*(3), 223–248.

Graham, S., & Healey, P. 1999. Relational Concepts of Space and Place: Issues for Planning Theory and Practice. *European Planning Studies, 7*(5) (October), 623.

Gunderson, L. H., & Holling, C. S. 2001. *Panarchy: Understanding Transformations in Human and Natural Systems*. Washington, DC and Covelo, London: Island Press.

Gustafsson, B. 1998. Scope and Limits of the Market Mechanism in Environmental Management. *Ecological Economics, 24*(2–3), 259–274.

Hajer, M. A. 1997. *The Politics of Environmental Discourse: Ecological Modernization and the Policy Process*. Oxford: Clarendon Press.

Hajer, M. A., & Wagenaar, H. 2003. *Deliberative Policy Analysis: Understanding Governance in the Network Society*. Cambridge, UK; New York: Cambridge University Press.

Hansen, J. M. 1991. *Gaining Access: Congress and the Farm Lobby, 1919–1981*. Chicago, IL: University of Chicago Press.

Hardy, S. D. 2010. Governments, Group Membership, and Watershed Partnerships. *Society & Natural Resources, 23*(7), 587–603.

Hartigan, J., Quasebarth, T., & Southerland, E. 1983. Calibration of NPS Model Loading Factors. *Journal of Environmental Engineering, 109*(6), 1259–1272.

Harvey, D. 1996. *Justice, Nature, and the Geography of Difference*. Cambridge, MA: Blackwell.

Heal, G. M. 2000. *Nature and the Marketplace: Capturing the Value of Ecosystem Services*. Washington, DC: Island Press.

Heal, G., Daily, G. C., Ehrlich, P. R., & Salzman, J. 2001. Protecting Natural Capital through Ecosystem Service Districts. *Stanford Environmental Law Journal, 20*, 333.

Heynen, N. 2007. *Neoliberal Environments: False Promises and Unnatural Consequences*. London, UK: Routledge.

Higgs, E. S. 1997. What Is Good Ecological Restoration? *Conservation Biology, 11*(2), 338–348.

Hilderbrand, R. H., Watts, A. C., & Randle, A. M. 2005. The Myths of Restoration Ecology. *Ecology and Society, 10*(1), 19.

Hirt, Paul W. 1996. *A Conspiracy of Optimism: Management of the National Forests Since World War Two*. Lincoln: University of Nebraska Press.

Hood, C. 1986. *The Tools of Government* (1st American ed.). Chatham, NJ: Chatham House.

Hood, C. 1991. A Public Management for All Seasons? *Public Administration, 69*(1), 3–19.

Hott, L., & Garey, D. 1989. *The Wilderness Idea: John Muir, Gifford Pinchot and the First Great Battle for Wilderness*. Video documentary. Santa Monica, CA: Direct Cinema Limited.

Houck, O. A. 2002. *The Clean Water Act TMDL Program: Law, Policy, and Implementation*. 2nd ed. Environmental Law Institute.

Hough, P., & Robertson, M. 2008. Mitigation under Section 404 of the Clean Water Act: Where It Comes From, What It Means. *Wetlands Ecology and Management, 17*(1) 15–33.

Innes, J. E., & Booher, D. E. 2010. *Planning with Complexity: An Introduction to Collaborative Rationality for Public Policy*. 1st ed. New York and London: Routledge.

Institute for Water Resources. 2015. "The Mitigation Rule Retrospective: A Review of the 2008 Regulations Governing Compensatory Mitigation for Losses of Aquatic Resources." 2015-R-03. Alexandria, VA: U.S. Army Corps of Engineers and Environmental Protection Agency.

Jasanoff, S. 1990. *The Fifth Branch: Science Advisors as Policymakers*. Cambridge, MA: Harvard University Press.

Jasanoff, S. 1995. *Science at the Bar: Law, Science, and Technology in America*. Cambridge, MA: Harvard University Press.

Jasanoff, S. 2004. *States of Knowledge: the Co-production of Science and Social Order*. London; New York: Routledge.

Jasanoff, S. 2005. *Designs on Nature: Science and Democracy in Europe and the United States*. Princeton, NJ: Princeton University Press.

Jasanoff, S., & Kim, S.-H. 2009. Containing the Atom: Sociotechnical Imaginaries and Nuclear Power in the United States and South Korea. *Minerva, 47*(2), 119–146.

Jerrick, N. 2001. *The Willamette Restoration Strategy*. Portland, OR: Willamette Restoration Initiative. Retrieved from http://oe.oregonexplorer.info/WillametteExplorerClassic/publications/pdf/WRS_OVER.pdf. Last accessed on 07/13/2013.

Johnson, N., White, A., & Perrot-Maitre, D. 2000. *Developing Markets for Water Services from Forests: Issues and Lessons for Innovators*. Washington, DC: Forest Trends. Retrieved from http://www.forest-trends.org/documents/files/doc_133.pdf. Last accessed on 07/13/2013.

Jones, C., Bacon, L., Kieser, M., & Sheridan, D. 2006. *Water-Quality Trading: A Guide for the Wastewater Community*. New York: McGraw-Hill.

Jones, C., Branosky, E., Selman, M. & Perez, M. 2010. *How Nutrient Trading Could Help Restore the Chesapeake Bay*. World Resources Institute Working Paper, Washington DC.

Kaiser, E. J., & Godschalk, D. R. 1995. Twentieth Century Land Use Planning: A Stalwart Family Tree. *Journal of the American Planning Association, 61*(3), 365–385.

Kane, A., & Prendergrass, J. 2007. *Bay Bank Baseline Analysis: Regulatory Drivers of Ecosystem Markets in the Chesapeake Bay.* August. Washington, DC: Environmental Law Institute.

Keller, A. A., Chen, X., Fox, J., Fulda, M., Dorsey, R., Seapy, B., Glenday, J., & Bray, E. 2014. Attenuation Coefficients for Water Quality Trading. *Environmental Science & Technology, 48*(12), 6788–6794.

Kieser, M., & Logue, C. 2015. *Advances in Water Quality Trading as a Flexible Compliance Tool: A Special Publication.* Alexandria, VA: Water Environment Foundation.

King, D. 2005. Crunch Time for Water Quality Trading. *Choices, 20*(1), 71–75.

King, D., & Herbert, L. 1997. The Fungibility of Wetlands. *National Wetlands Newsletter, 19*(5), 10–13.

King, D., & Kuch, P. 2003. Will Nutrient Credit Trading Ever Work? An Assessment of Supply and Demand and Institutional Obstacles. *Environmental Law Reporter, 33*, 10352–10368.

Kroeger, T., & Casey, F. 2007. An Assessment of Market-Based Approaches to Providing Ecosystem Services on Agricultural Lands. *Ecological Economics, 64*(2), 321–332.

Krugman, P. 2010, April 7. Climate Change—Building a Green Economy. *The New York Times.* Retrieved from http://www.nytimes.com/2010/04/11/magazine/11Economy-t.html.

Latour, B. 1987. *Science in Action: How to Follow Scientists and Engineers through Society.* Cambridge, MA: Harvard University Press.

Layzer, J. A. 2008. *Natural Experiments: Ecosystem-based Management and the Environment.* Cambridge, MA: MIT Press.

Layzer, J. A. 2012. *Open for Business: Conservatives' Opposition to Environmental Regulation.* Cambridge, MA: MIT Press.

Leach, W., & Pelkey, N. 2001. Making Watershed Partnerships Work: A Review of the Empirical Literature. *Journal of Water Resources Planning and Management, 127*(6), 378–385.

Lemos, M. C., & Agrawal, A. 2006. Environmental Governance. *Annual Review of Environment and Resources, 31*(1), 297–325.

Leopold, A. 1949. *A Sand County Almanac.* New York: Oxford University Press.

Levin, P. S., Fogarty, M. J., Murawski, S. A., & Fluharty, D. 2009. Integrated Ecosystem Assessments: Developing the Scientific Basis for Ecosystem-Based Management of the Ocean. *PLoS Biol, 7*(1).

Lewis, W. M. 2001. *Wetlands Explained: Wetland Science, Policy, and Politics in America.* Oxford; New York: Oxford University Press.

Linker, L. C., Shenk, G. W., Wang, P., Hopkins, K. J., & Pokharel, S. 2002. A Short History of Chesapeake Bay Modeling and the Next Generation of Watershed and Estuarine Models. *Proceedings of the Water Environment Federation, 2002*(2), 569–582.

Linkous, E. R., & Chapin, T. S. 2014. TDR Program Performance in Florida. *Journal of the American Planning Association, 80*(3), 253–267.

Lynch, K. 1981. *A Theory of Good City Form.* Cambridge, MA: MIT Press.

Mackintosh, B. 1987. The National Park Service Moves into Historical Interpretation. *The Public Historian, 9*(2), 51–63.

Madsen, B., Carroll, N., & Brands, K. M. 2010. "State of Biodiversity Markets: Offset and Compensation Programs Worldwide." Washington, DC: Ecosystem Marketplace. http://www.forest-trends.org/documents/files/doc_2388.pdf.

Madsen, B., Kandy, D., & Bennett, G. 2011. "2011 Update: State of Biodiversity Markets. Offset and Compensation Programs Worldwide." Washington, DC: Forest Trends. http://www.forest-trends.org/documents/files/doc_2848.pdf.

Majanen, T., Friedman, R., & Milder, J. 2011. "Innovations in Market-based Watershed Conservation in the United States: Payments for Watershed Services for Agricultural and Forest Landowners." Ecoagriculture Partners. http://www.ecoagriculture.org/documents/files/doc_362.pdf.

Malinowski, B. 1922. *Argonauts of the Western Pacific: An Account of Native Enterprise and Adventure in the Archipelagoes of Melanesian New Guinea*. London; New York: G. Routledge; E. P. Dutton.

Mann, C., & Absher, J. D. 2014. Adjusting Policy to Institutional, Cultural and Biophysical Context Conditions: The Case of Conservation Banking in California. *Land Use Policy, 36*(January), 73–82.

Mansfield, B. 2009. *Privatization: Property and the Remaking of Nature–Society Relations*. Malden, MA: Wiley.

March, J. G. 1994. *Primer on Decision Making: How Decisions Happen*. 1st ed. New York: Free Press.

March, J. G. 2006. "The Logic of Appropriateness." In *The Oxford Handbook of Public Policy*, edited by Michael Moran, Martin Rein and Robert E. Goodin, 689–708. Oxford [England]: Oxford University Press.

March, J. G., & Olsen, J. P. 1989. *Rediscovering Institutions: The Organizational Basis of Politics*. New York: Free Press.

Marcus, G. E. 1995. *Technoscientific Imaginaries: Conversations, Profiles, and Memoirs*. Chicago, IL: University of Chicago Press.

Marcus, G. E. 1998. *Ethnography through Thick and Thin*. Princeton, NJ: Princeton University Press.

Marcus, G. E., & Fischer, M. 1986. *Anthropology as Cultural Critique: An Experimental Moment in the Human Sciences*. Chicago: University of Chicago Press.

Marshall, E., & Weinberg, M. 2012. "Baselines in Environmental Markets: Tradeoffs Between Cost and Additionality." EB-18. Economic Brief. Washington, DC: U.S. Department of Agriculture Economic Research Service. http://www.ers.usda.gov/media/281180/eb18_1_.pdf.

Mayrand, K., & Paquin, M. 2004. "Payments for Environmental Services: A Survey and Assessment of Current Schemes." Montreal: Unisfera International Centre.

McCormick, J. 1995. *The Global Environmental Movement*. Chichester [England]; New York: Wiley.

McGinnis, S. L. 2001. Watershed-based Pollution Trading Development and Current Trading Programs. *Environmental Engineering and Policy, 2*(3), 161–170.

McHarg, I. L. 1969. *Design with Nature*. Garden City, NY: Published for the American Museum of Natural History [by] the Natural History Press.

Molnar, J. L., & Kubiszewski, I. 2012. Managing Natural Wealth: Research and Implementation of Ecosystem Services in the United States and Canada. *Ecosystem Services, 2*(December), 45–55.

Moore, R. 2014. *The Role of Trading in Achieving Water Quality Objectives*. Washington, DC. Available at: http://transportation.house.gov/calendar/eventsingle.aspx?EventID= 373351].

Moore, R. H., Parker, J. S., & Weaver, M. 2008. Agricultural Sustainability, Water Pollution, and Governmental Regulations: Lessons from the Sugar Creek Farmers in Ohio. *Culture & Agriculture, 30*(1–2), 3–16.

Morris, R. K. A., Alonso, I., Jefferson, R. G., & Kirby, K. J. 2006. The Creation of Compensatory Habitat—Can It Secure Sustainable Development? *Journal for Nature Conservation, 14*(2), 106–116.

Muir, J. 1912. *The Yosemite*. New York: The Century Company.

Müller, F., & Burkhard, B. 2012. The Indicator Side of Ecosystem Services. *Ecosystem Services, 1*(1) (July), 26–30.

Munda, G. 1997. Environmental Economics, Ecological Economics, and the Concept of Sustainable Development. *Environmental Values, 6*(2) (May), 213–233.

Muradian, R., Corbera, E., Pascual, U., Kosoy, N., & May, P. H. 2010. Reconciling Theory and Practice: An Alternative Conceptual Framework for Understanding Payments for Environmental Services. *Ecological Economics, 69*(6), 1202–1208.

Naess, A. 1990. *Ecology, Community and Lifestyle: Outline of an Ecosophy.* Cambridge and New York: Cambridge University Press.

National Research Council (U.S.) 2001. *Compensating for Wetland Losses Under the Clean Water Act.* Washington, DC: National Academies Press.

National Research Council (U.S.). 2005. *Valuing Ecosystem Services: Toward Better Environmental Decision-Making.* Washington, DC: National Academies Press.

National Research Council. 2011. *Achieving Nutrient and Sediment Reduction Goals in the Chesapeake Bay: An Evaluation of Program Strategies and Implementation.* Washington, DC: National Academies Press.

Newburn, D. A., & Woodward, R. T. 2012. An Ex Post Evaluation of Ohio's Great Miami Water Quality Trading Program. *Journal of the American Water Resources Association, 48*(1), 156–169.

Norgaard, R. B. 2010. Ecosystem Services: From Eye-Opening Metaphor to Complexity Blinder. *Ecological Economics, 69*(6), 1219–1227.

ORDEQ. 2011. *NPDES Permit Evaluation Report and Fact Sheet.* Medford, OR. Retrieved from http://northwestenvironmentaladvocates.org/nweafiles/WQ_Trading/download_1/MedfordEVAL%5b1%5d.pdf.

O'Leary, R., & Bingham, L. 2003. *The Promise and Performance of Environmental Conflict Resolution.* Washington, DC: Resources for the Future.

Osei, E., Gassman, P. W., Jones, R. D., Pratt, S. J., Hauck, L. M., Beran, L. J., Rosenthal, W. D., & Williams, J. R. 2000. Economic and Environmental Impacts of Alternative Practices on Dairy Farms in an Agricultural Watershed. *Journal of Soil and Water Conservation, 55*(4) (October 1), 466–472.

Ozawa, C. P. 1991. *Recasting Science: Consensual Procedures in Public Policy Making.* Boulder, CO: Westview Press.

Paehlke, R., & Torgerson, D. 2005. *Managing Leviathan: Environmental Politics and the Administrative State.* Peterborough, Ontario, Canada; Lewiston, NY: Broadview Press.

Polanyi, M. 1962. The Republic of Science: Its Political and Economic Theory. *Minerva, 1*(1), 54–73.

Popular Science. 2008, February 8. America's 50 Greenest Cities. Retrieved April 30, 2013, from http://www.popsci.com/environment/article/2008-02/americas-50-greenest-cities.

Porter, T. M. 1995. *Trust in Numbers: The Pursuit of Objectivity in Science and Public Life.* Princeton, NJ: Princeton University Press.

Pressman, J. L. 1984. *Implementation: How Great Expectations in Washington Are Dashed in Oakland: Or, Why It's Amazing that Federal Programs Work at All, this Being a Saga of the Economic Development Administration as Told by Two Sympathetic Observers Who Seek to Build Morals on a Foundation of Ruined Hopes.* 3rd ed. Berkeley: University of California Press.

Primozich, D. 2008. *Developing the Willamette Ecosystem Marketplace.* Willamette Partnership. Retrieved from http://willamettepartnership.org/publications/MarketplacePubs/developing-the- willamette-ecosystem-marketplace.pdf.

Rabalais, Nancy N., Turner, R. E., & Wiseman, W. J. 2002. Gulf of Mexico Hypoxia, A.K.A. "The Dead Zone." *Annual Review of Ecology and Systematics, 33*(1), 235–263.

Richter, A., & Kolmes, S. A. 2005. Maximum Temperature Limits for Chinook, Coho, and Chum Salmon, and Steelhead Trout in the Pacific Northwest. *Reviews in Fisheries Science, 13*(1), 23–49.

Reed, M. S., Graves, A., Dandy, N., Posthumus, H., Hubacek, K., Morris, J., Prell, C., Quinn, C. H., & Stringer, L. C. 2009. Who's in and Why? A Typology of Stakeholder Analysis Methods for Natural Resource Management. *Journal of Environmental Management, 90*(5), 1933–1949.

Robertson, M. 2006. The Nature that Capital Can See: Science, State, and Market in the Commodification of Ecosystem Services. *Environment and Planning D: Society and Space, 24*(3), 367–387.

Robertson, M. 2007. Discovering Price in All the Wrong Places: The Work of Commodity Definition and Price under Neoliberal Environmental Policy. *Antipode, 39*(3), 500–526.

Robertson, M. 2009. The Work of Wetland Credit Markets: Two Cases in Entrepreneurial Wetland Banking. *Wetlands Ecology and Management, 17*(1) (February), 35–51.

Robertson, M. 2012. Measurement and Alienation: Making a World of Ecosystem Services. *Transactions of the Institute of British Geographers, 37*(3), 386–401.

Robertson, M., BenDor, T., Lave, R., Riggsbee, A. & Doyle, M. 2014. "Stacking Ecosystem Services." *Frontiers in Ecology and the Environment, 12*(3), 186–193.

Robertson, M., & Hayden, N. 2008. Evaluation of a Market in Wetland Credits: Entrepreneurial Banking in Chicago. *Conservation Biology, 22*(3), 636–646.

Rockloff, S. F., & Lockie, S. 2004. Participatory Tools for Coastal Zone Management: Use of Stakeholder Analysis and Social Mapping in Australia. *Journal of Coastal Conservation, 10*(1), 81–92.

Ruhl, J. B. 2010. Ecosystem Services and the Clean Water Act: Strategies for Fitting New Science into Old Laws. *SSRN eLibrary.* Retrieved from http://papers.ssrn.com/sol3/papers.cfm?abstract_id=1666246.

Ruhl, J. B., Kraft, S. E., & Lant, C. L. 2007. *The Law and Policy of Ecosystem Services.* Washington, DC: Island Press.

Ruhl, J. B., & Salzman, J. 2006. The Effects of Wetland Mitigation Banking on People. *National Wetlands Newsletter, 28*(2), 9–14.

Sabatier, P. A. 2005. *Swimming Upstream: Collaborative Approaches to Watershed Management.* Cambridge, MA: MIT Press.

Sachs, A. 2013. *Arcadian America: The Death and Life of an Environmental Tradition.* New Haven, CT and London: Yale University Press.

Sagoff, M. 2002. On the Value of Natural Ecosystems: The Catskills Parable. *Politics and the Life Sciences, 21*(1), 19–25.

Salamon, L. M., & Elliott, O. V. (Eds.) 2002. *The Tools of Government: A Guide to the New Governance.* Oxford: Oxford University Press.

Salamon, L. M., & Lund, M. S. (Eds.) 1988. *Beyond Privatization: The Tools of Government Action.* Washington, DC: Urban Institute Press.

Salzman, J. 1997. Valuing Ecosystem Services. *Ecology LQ, 24*, 887.

Salzman, J. 2005a. Creating Markets for Ecosystem Services: Notes from the Field. *NYU Law Review, 80*, 870–961.

Salzman, J. 2005b. The Promise and Perils of Payments for Ecosystem Services. *International Journal of Innovation and Sustainable Development*, *1*(1), 5–20.

Salzman, J., & Ruhl, J. B. 2000. Currencies and the Commodification of Environmental Law. *Stanford Law Review*, *53*, 607–694.

Sample, V. A. (n.d.). *The Pinchot Institute at 50: A Brief History*. Pinchot Institute for Conservation.

Sandel, M. J. 2012. *What Money Can't Buy: The Moral Limits of Markets*. New York: Farrar, Straus and Giroux.

Sanneman, C., Culliney, S., & Cochran, B. 2014. "Verification in Markets for Water Quality & Habitat." U.S. Department of Agriculture Office of Environmental Markets. Portland, OR: Willamette Partnership. http://willamettepartnership.org/wp-content/uploads/2014/10/Verification-in-Markets-for-WQ-and-Habitat_2014-10-01.pdf.

Scarlett, L., & Boyd, J. 2011. *Ecosystem Services: Quantification, Policy Applications, and Current Federal Capabilities* (Discussion Paper No. RFF DP 11–13). Washington, DC: Resources for the Future. Retrieved from http://www.rff.org/RFF/Documents/RFF-DP-11–13.pdf. Last accessed on 07/17/2013.

Schenk, T. 2007. "Conflict Assessment: A Review of the State of Practice." Consensus Building Institute. http://cbuilding.org/sites/cbi.drupalconnect.com/files/ConflictAssessmentSummary_Schenk.pdf.

Schnabel, R., & Gburek, W. 1985. Discussion of "Calibration of NPS Model Loading Factors" by John P. Hartigan, Thomas F. Quasebarth, and Elizabeth Southerland (December 1983). *Journal of Environmental Engineering*, *111*(1), 103–107.

Selman, M., Sprague, E., Walker, S., & Kittler, B. 2010. "Nutrient Trading in the Chesapeake Region: An Analysis of Supply and Demand." Washington, DC: Pinchot Institute for Conservation. http://www.pinchot.org/uploads/download?fileId=658.

Seidman, H. 1998. *Politics, Position, and Power: The Dynamics of Federal Organization*. New York: Oxford University Press.

Shabman, L., Stephenson, K., & Shobe, W. 2002. Trading Programs for Environmental Management: Reflections on the Air and Water Experiences. *Environmental Practice*, *4*(3), 153–162.

Shandas, V., & Messer, W. B. 2008. Fostering Green Communities Through Civic Engagement: Community-based Environmental Stewardship in the Portland Area. *Journal of the American Planning Association*, *74*(4), 408–418.

Sierra Club. 2002. "The Bush Administration's Record on the 30th Anniversary of the Clean Water Act." Sierra Club.

Simpson, T., & Weammart, S. 2009. *Developing Best Management Practice Definitions and Effectiveness Estimates for Nitrogen, Phosphorous and Sediment in the Chesapeake Bay Watershed*. University of Maryland Mid-Atlantic Water Program. Retrieved from http://archive.chesapeakebay.net/pubs/BMP_ASSESSMENT_REPORT.pdf.

Smith, A. [1776]. Edited by Skinner, A. S. 1999. *The Wealth of Nations*. London; New York: Penguin.

Smith, D., Ammann, A., Bartoldus, C. C., & Brinson, M. 1995. "An Approach for Assessing Wetland Functions Using HGM Classification, Reference Wetlands, and Functional Indices." Technical Report WRP-DE-9. Wetlands Research Program Technical Report. Vicksburg, MS: U.S. Army Engineer Waterways Experiment Station.

Smith, N. 2007. Nature as Accumulation Strategy. *Socialist Register*, *2007*, 16.

Sommerville, M. M., Jones, J. P. G., & Milner-Gulland, E. J. 2009. A Revised Conceptual Framework for Payments for Environmental Services. *Ecology and Society*, *14*(2), 34.

Spirn, A. W. 1984. *The Granite Garden: Urban Nature and Human Design.* New York: Basic Books.

Sprague, E., Burke, D., Claggett, S., & Todd, A. 2006. *The State of the Chesapeake Forest.* Washington, DC: The Conservation Fund. Retrieved from http://www.na.fs.fed.us/watershed/pdf/socf/Full%20Report.pdf.

Stanton, T., Echavarria, M., Hamilton, K., & Ott, C. 2010. *State of Watershed Payments: An Emerging Marketplace.* Ecosystem Marketplace. Retrieved from http://www.forest-trends.org/documents/files/doc_2438.pdf. Last accessed on 07/17/2013.

Star, S. L., & Griesemer, J. R. 1989. Institutional Ecology, Translations, and Boundary Objects: Amateurs and Professionals in Berkeley's Museum of Vertebrate Zoology, 1907–39. *Social Studies of Science, 19*(3), 387–420.

Stavins, R. N. 1998. What Can We Learn From the Grand Policy Experiment? Lessons from SO2 Allowance Trading. *Journal of Economic Perspectives, 12*(3), 69–88.

Stavins, R. N. 2003. "Chapter 9 Experience with Market-based Environmental Policy Instruments." In *Handbook of Environmental Economics,* edited by Karl-Göran Mäler and Jeffrey R. Vincent, Volume 1. 355–435. Elsevier.

Steinzor, R. I., & Jones, S. 2013. Collaborating to Nowhere: The Imperative of Government Accountability for Restoring the Chesapeake Bay. Retrieved from http://digitalcommons.law.umaryland.edu/fac_pubs/1302/.

Stephenson, K., & Bosch, D. J. 2003. "Nonpoint Source and Carbon Sequestration Credit Trading: What Can the Two Learn from Each Other?" 2003 Annual Meeting, July 27–30, Montreal, Canada. http://www.envtn.org/uploads/stephenson_NPScarbon_2003.pdf.

Stephenson, K., Kerns, W., & Shabman, L. 1995. *Market-based Strategies & Nutrient Trading: What You Need To Know.* Blacksburg: Virginia Tech.

Stephenson, K., & Shabman, L. 2011. Rhetoric and Reality of Water Quality Trading and the Potential for Market-like Reform. *JAWRA Journal of the American Water Resources Association, 47*(1), 15–28.

Stone, D. A. 2002. *Policy Paradox: The Art of Political Decision Making* (Rev. ed.). New York: Norton.

Susskind, L. 1985. The Siting Puzzle: Balancing Economic and Environmental Gains and Losses. *Environmental Impact Assessment Review, 5*(2) (June), 157–163.

Susskind, L., & Cruikshank, J. 1987. *Breaking the Impasse: Consensual Approaches to Resolving Public Disputes.* New York: Basic Books.

Susskind, L., McKearnan, S., & Thomas-Larmer, J. 1999. *The Consensus Building Handbook: A Comprehensive Guide to Reaching Agreement.* Thousand Oaks, CA: Sage.

Talberth, J., Jones, C., Perez, M., Selman, M., & Branosky, E. 2010. *How Baywide Nutrient Trading Could Benefit Maryland Farms.* Washington, DC: World Resources Institute.

Taylor, J. E. 1999. *Making Salmon: An Environmental History of the Northwest Fisheries Crisis.* University of Washington Press.

Thoreau, H. D. [1854] 2009. *Walden.* New York: Cosimo Classics.

Tietenberg, T. H. 1990. Economic Instruments for Environmental Regulation. *Oxford Review of Economic Policy, 6*(1), 17–33.

Tietenberg, T. H. 2006. *Emissions Trading: Principles and Practice.* Washington, DC: Resources for the Future.

Tietenberg, T. H., & Lewis, L. 2012. *Environmental & Natural Resource Economics.* Upper Saddle River, NJ: Pearson Education.

Tilman, D., Cassman, K. G., Matson, P. A., Naylor, R., & Polasky, S. 2002. Agricultural Sustainability and Intensive Production Practices. *Nature, 418*(6898), 671–677.

Troy, A., & Wilson, M. 2006. Mapping Ecosystem Services: Practical Challenges and Opportunities in Linking GIS and Value Transfer. *Ecological Economics, 60*(2), 435–449.

Van Der Pijl, K. 1998. *Transnational Classes and International Relations*. London and New York: Routledge.

Van Houtven, G., Loomis, R., Baker, J., Beach, R., & Casey, S. 2012. "Nutrient Credit Trading for the Chesapeake Bay: An Economic Study." Research Triangle Parl, NC: RTI International for the Chesapeake Bay Commission. http://www.chesbay.us/Publications/nutrient-trading-2012.pdf.

Vatn, A. 2010. An Institutional Analysis of Payments for Environmental Services. *Ecological Economics, 69*(6), 1245–1252.

Vileisis, Ann. 1999. *Discovering the Unknown Landscape: A History of America's Wetlands*. Washington, DC: Island Press.

Wellman, J. D. 1987. Foresters' Core Values and Cognitive Styles: Issues for Wildland Recreation Management and Policy. *Review of Policy Research, 7*(2), 395–403.

Westman, W. E. 1977. How Much Are Nature's Services Worth? *Science, 197*(4307), 960–964.

Wheeler, D. P., & Strock, J. M. 1995. "Official Policy on Conservation Banks." The California Resources Agency and California Environmental Protection Agency.

White, R. 2004. From Wilderness to Hybrid Landscapes: The Cultural Turn in Environmental History. *Historian, 66*(3), 557–564.

Whitworth, J. S. 2015. *Quantified: Redefining Conservation for the Next Economy*. Washington, DC: Island Press.

Willamette Partnership. 2009. Ecosystem Credit Accounting: Pilot General Crediting Protocol: Willamette Basin Version 1.1.

Willamette Partnership, Pinchot Institute for Conservation, and World Resources Institute. 2012. "In It Together: A How-To Reference to Building Point-Nonpoint Water Quality Trading Programs." Hillsboro, OR: United States Department of Agriculture.

Willamette Partnership, World Resources Institute, and the National Network on Water Quality Trading. 2015. "Building a Water Quality Trading Program: Options and Considerations." National Network on Water Quality Trading.

Williamson, O. E. 1981. The Economics of Organization: The Transaction Cost Approach. *American Journal of Sociology* (November), 548–577.

Wilson, J. Q. 1989. *Bureaucracy: What Government Agencies Do and Why They Do It*. New York: Basic Books.

Wilson, M. A., & Howarth, R. B. 2002. Discourse-based Valuation of Ecosystem Services: Establishing Fair Outcomes through Group Deliberation. *Ecological Economics, 41*(3), 431–443.

Wise, S. 2008. Green Infrastructure Rising. *Planning, 74*(8), 14–19.

Weber, M. 1968. *From Max Weber: Essays in Sociology*. H. H. Gerth, & C. W. Mills (Eds.). New York: Oxford University Press.

Womble, P., & Doyle, M. 2012. The Geography of Trading Ecosystem Services: A Case Study of Wetland and Stream Compensatory Mitigation Markets. *Harvard Environmental Law Review, 36*, 229–296.

Wondolleck, J. M., & Yaffee, S. L. 2000. *Making Collaboration Work : Lessons from Innovation in Natural Resource Management*. Washington, DC: Island Press.

Wu, J-J, & Babcock, B. A. 1996. Contract Design for the Purchase of Environmental Goods from Agriculture. *American Journal of Agricultural Economics, 78*(4), 935–945.

Wunder, S. 2005. Payments for Environmental Services: Some Nuts and Bolts. CIFOR Occasional Paper 42.

INDEX